THE HISTORY
OF CLASSICAL PHYSICS

BIBLIOGRAPHIES OF THE HISTORY
OF SCIENCE AND TECHNOLOGY
(Vol. 8)

GARLAND REFERENCE LIBRARY
OF THE HUMANITIES
(Vol. 444)

Bibliographies of the History of Science and Technology

Editors

Robert Multhauf, Smithsonian Institution, Washington, D.C.
Ellen Wells, Smithsonian Institution, Washington, D.C.

THE HISTORY
OF CLASSICAL PHYSICS
A Selected, Annotated Bibliography

R.W. Home
with the assistance of Mark J. Gittins

GARLAND PUBLISHING, INC. • NEW YORK & LONDON
1984

PHYSICS

Library of Congress Cataloging in Publication Data

Home, Roderick Weir.
  The history of classical physics.

  (Bibliographies of the history of science
and technology ; vol. 8) (Garland reference library of the
humanities ; vol. 444)
  Includes index.
  1. Physics—Bibliography. 2. Physics—History—
Bibliography. I. Gittins, Mark J. II. Title.
III. Series: Bibliographies of the history of science
and technology ; v. 8. IV. Series: Garland reference
library of the humanities ; v. 444.
Z7141.H65 1984 [QC21.2] 016.53 83-48276
ISBN 0-8240-9067-5 (alk. paper)

Printed on acid-free, 250-year paper
Manufactured in the United States of America

CONTENTS

GENERAL INTRODUCTION

This bibliography is one of a series designed to guide the reader into the history of science and technology. Anyone interested in any of the components of this vast subject area is part of our intended audience, not only the student, but also the scientist interested in the history of his own field (or faced with the necessity of writing an "historical introduction") and the historian, amateur or professional. The latter will not find the bibliographies "exhaustive," although in some fields he may find them the only existing bibliographies. He will in any case not find one of those endless lists in which the important is lumped with the trivial, but rather a "critical" bibliography, largely annotated, and indexed to lead the reader quickly to the most important (or only existing) literature.

Inasmuch as everyone treasures bibliographies, it is surprising how few there are in this field. George Sarton's *Guide to the History of Science* (Waltham, Mass., 1952; 316 pp.), Eugene S. Ferguson's *Bibliography of the History of Technology* (Cambridge, Mass., 1968; 347 pp.), François Russo's *Histoire des Sciences et des Techniques. Bibliographie* (Paris, 2nd ed., 1969; 214 pp.) are justifiably treasured but they are, of necessity, limited in their coverage and need to be updated.

For various reasons, mostly bad, the average scholar prefers adding to the literature to sorting it out. The editors are indebted to the scholars represented in this series for their willingness to expend the time and effort required to pursue the latter objective. Our aim, and that of the publisher, has been to give the series enough uniformity to give some consistency to the series, but otherwise to leave the format and contents to the author/compiler. We have urged that introductions be used for essays on "the state of the field," and that selectivity be exercised to limit the length of each volume. Since the historical literature ranged from very large (e.g., medicine) to very small (e.g., chemical technology), some bibliographies will be limited to truly important writings while others will include modest "contributions" and even primary sources. The problem is intelligible guidance into a particular field—or subfield—and its solution is largely left to the author/compiler. In general, topical volumes (e.g., chemistry) will deal with the subject since about 1700, leaving earlier literature to area or chronological volumes (e.g., medieval science); but here, too, the volumes will vary according to the judgment of the author. The volumes are international (except for two, *Science and Technology in the United States* and *Science and Technology in Eastern Asia*) but the literature covered depends, of course, on the linguistic equipment of the author and his access to "exotic" literatures.

Robert Multhauf
Ellen Wells

Smithsonian Institution
Washington, D.C.

INTRODUCTION

In compiling this bibliography, "classical physics" has been taken to embrace the period between approximately 1700 and 1900. So far as the earlier cut-off date is concerned, the two major figures of Newton and Leibniz have been taken to mark the point of transition from the preceding period of the Scientific Revolution. Material relating primarily to the dissemination of their ideas has been included even when, as in one or two instances, this deals primarily with events of the 1690s. On the other hand, studies dealing with the works of Newton and Leibniz themselves have been left for the volume on the Scientific Revolution even when, as in the case of the famous Leibniz–Clarke correspondence, they date from as late as 1716 or 1717. In other words, Newton and Leibniz themselves have been taken to belong to the period of the Scientific Revolution, their influence to belong to the subsequent era of classical physics.

The later cut-off point has been established with the commonly drawn distinction between "classical" and "modern" physics in mind. Thus, items devoted to X-rays or radioactivity have been omitted even where these deal exclusively with the period prior to 1900, whereas studies of "classical" aether-based electromagnetic theory have been included even when the events with which they are concerned occurred as late as 1910 or 1912. Items dealing with relativity or quantum theory have been included only if they contain significant discussions of pre-relativity or pre-quantum ideas respectively. In short, items have been included that deal to a significant extent with the "classical" background to one or another aspect of 20th-century physics but omitted if they are concerned exclusively or almost entirely with the "modern" aspects of the subject.

It also had to be decided what was to be counted as "physics," and here difficulties sometimes arose not just because, inevitably, some studies dealt with topics close to the margins of the discipline but also because the denotation of the word "physics" itself changed considerably during the period under review. In 1700, "physics" (or "natural philosophy," its closest English equivalent) was still widely understood in one of its traditional Aristotelian usages as the science of "nature" in general. Physical treatises of the time typically included substantial sections on anatomy, botany, chemistry, and "the system of the world" (that is, the theory of the heavens). Only very gradually during the course of the eighteenth century did the subject come to be restricted to the much narrower range of topics that it embraces today. (Cf. item 452).

In the end, the decision as to what to include or to exclude was chiefly determined by the structure of the broader series of bibliographies within which this work is being published. Within the series as a whole, lines of demarcation between the different volumes have, for various good reasons, been drawn on the basis of the present-day rather than the historical boundaries between scientific disciplines, and we have felt bound to conform to this arrangement as nearly as we could. Hence, for example, we have taken the caloric theory of gases to fall squarely within our purview, even though its originator, Lavoisier, undoubtedly saw it as a primarily *chemical* theory. By contrast, we have omitted almost everything to do with astronomy, whether devoted to astrophysical matters or to the motions of the planets, and likewise material to do with the science of microscopy except for those items that deal with the design and construction of the microscopes themselves. Similarly, we have omitted material dealing with questions of meteorology, geophysics—including terrestrial magnetism—and electrical technology because other volumes in the series will be devoted to those subjects.

Subject, however, to these constraints imposed by the structure of the series as a whole, we have interpreted the word "physics" in a fairly broad sense. In particular, we have included a substantial body of literature on the theory of matter, some of which might be felt these days to belong rather to metaphysics on the one hand or to chemistry on the other. We have also included, for example, a number of general studies of eighteenth-century Newtonianism and its rivals, even when, as in a couple of papers on the ideas of the Hutchinsonians, most of the issues being discussed are in fact fairly remote from the domain of present-day physics.

This is primarily a bibliography of current writing on the history of physics, not a listing of sources. In fact, only two classes of primary sources have been included. In the belief that editions of collected works of individual physicists, though unfortunately not always as reliable as one might wish, are often extremely convenient to use, we have listed in Chapter III as many relevant publications of this kind as have come to our attention. In addition, particular editions of individual works have been listed when their editorial apparatus includes significant amounts of historical commentary.

As indicated in the General Introduction, our listings are selective rather than comprehensive. Certain classes of material have been excluded altogether, others are represented only patchily. Thus, we have omitted "popular," undocumented items altogether except in a very few special cases. We have tended to omit preliminary papers of the "work-in-progress" kind in favour of fuller, "final" versions published by the same author. Older works have not been included if in our judgment they have been entirely superseded by more recent studies. We have omitted theses and doctoral dissertations if these or substantial extracts from them have been published separately. Furthermore, we have not made any systematic effort to seek out other, unpublished theses—by and large, we have listed only those that came to our attention through being cited somewhere in a footnote.

Institutional histories have posed a particular problem because there is such a vast number of them, ranging from substantial histories of major scientific institutions to brief accounts of minor universities that have something to say in passing about the local department of physics. We have adopted a policy of including standard histories of the major eighteenth- and nineteenth-century scientific academies whenever such works exist, but those of lesser institutions only when they devote particular attention to matters of physics. We have also omitted most histories of individual colleges or universities and, again, included only studies that concentrate specifically on matters relating to our subject.

Biographies have also been treated in a highly selective manner, especially when (as in the case of Faraday, for example) there is a multiplicity of "popular" accounts as well as more scholarly studies. Our policy has been to omit items that are clearly derivative or superseded by more recent work and to include, in addition to old-style but often still valuable "life and letters" compilations whenever these exist, only those more recent studies that contribute in one way or another to serious historical debate. Nor have we included obituary notices or formal biographical memoirs prepared shortly after a person's death though we have noted (item 1a) a useful index to English-language writings of this kind. In many cases such articles are now superseded by the corresponding entries in that indispensable reference tool for anyone interested in any aspect of the history of science, the *Dictionary of Scientific Biography* (item 7). Where necessary this can be supplemented by the relevant national dictionaries of biography (none of which, however, we have felt it necessary to include in this bibliography) and, for at least an outline of a person's career together with a list of his or her publications, Poggendorff's invaluable compilation (item 15).

Though we have included items written in a variety of European languages (we are incompetent to survey contributions in non-European languages but believe that our selection has not significantly suffered thereby) there is an undoubted preponderance of material published in English. To some extent this is a result of deliberate editorial policy: for example, whenever an English translation exists of a work first published in some other language, we have listed the translation rather than the original edition, while conversely, we have not listed translations into other languages of works first published in English. In part, however, it also reflects the present domination of the field by Anglo-Saxon scholarship, allied with an increasing tendency for historians of science even from non-English speaking parts of the world to publish their major writings in English. We do not believe that recent work in other western European languages is seriously underrepresented in our listings. (Work published in eastern European languages—Russian, Polish, Czech, and so on—may be a different matter, as may works published in the German Democratic Republic.)

A list of the journals consulted in compiling this bibliography and the abbreviated titles used in citing them is provided on pages xvii–xix. Journals that were scanned systematically are marked with an asterisk. In surveying both journals and monographs, we relied for the most part on the very extensive holdings of the Baillieu Library, University of Melbourne, supplemented here and there by those of the University Library, Cambridge, and by the use of inter-library loan facilities. The effective cut-off date for inclusion of citations was 6 December 1982. To be included, however, material had to reach Melbourne by that date; and since surface mail to Australia from either Europe or North America often takes three months to reach its destination, a number of items may well have been omitted that were actually in print some time before then.

We have devised our own fairly simple classification scheme for our bibliography. The primary division is chronological and reflects what we believe were significant points of transition in the history of physics itself. Except in mechanics and certain parts of optics, physics remained largely qualitative and non-mathematical throughout most of the eighteenth century. In the last few years of the century, however, and in the first couple of decades of the next, mathematics began at last to be brought successfully to bear on various other areas of physical inquiry such as the conduction of heat, the theory of capillarity, wave optics, and the theory of electricity and magnetism. This development occurred chiefly in France and is linked particularly with such famous names as Coulomb, Laplace, Biot, Fourier, Poisson, Fresnel, and Ampère. During the same period we find a number of discoveries and new (or newly renovated) ideas emerging—for example, the Voltaic pile, the polarization of light by reflection, the magnetic effect of an electric current, *Naturphilosophie*, the wave theory of light, the atomic theory—and also the emergence of new institutional forms such as the reconstructed Académie des Sciences in Paris, the Ecole Polytechnique, and, in Berlin, the first explicitly research-oriented university. This transitional period ended around 1820. It could perhaps be argued that a further transition occurred in the 1850s associated with the rise of thermodynamics and the energy concept, but we have chosen instead to treat the remaining period from 1820 to 1900 as a unity. Furthermore, we have sought by means of extensive cross-referencing to link our transitional period, 1790–1820, with the preceding and following periods.

The two major chronological divisions (but not the transitional period, which has been treated rather differently) have been further divided according to subject matter, including sections on institutions and instrumentation, as well as on each of the usual branches of physical inquiry. In Chapter IV a similar classification has been imposed on works spanning more than one of our chronological divisions. In the main, however, these more general works have not been cross-referenced in the listings for the separate chronological periods. Hence to find out what has been published on, say, eighteenth-century optics, one needs to consult not just the relevant section in the appropriate chronological division (in this case, Section f of Chapter V) but also the section on Light in Chapter IV.

Inevitably, within each chronological division there is a substantial number of items that span two or more of the subdivisions or that for other reasons could not easily be placed into a single category. Chapters IV, V and VII all, therefore, contain sizeable sections of unclassified "general" items in addition to the items listed in the more specific subdivisions already described. While some cross-referencing of this material has been attempted, we expect that in many cases the index will provide the most convenient means of access to it. Within each subdivision, entries are listed alphabetically by author.

The index is one primarily of names. Where the person named is the author of the item in question, the entry is marked with an asterisk. Subject headings are included for a few general topics ("Newtonianism," for example) that are widely used but are not otherwise differentiated in our classification.

The field covered by this bibliography is one that has received much more concentrated attention in recent years than it did from previous generations of historians. As a result, most of the items listed are less than fifteen years old, and many of these have appeared within the past five years. Furthermore, there is no sign of slackening in the rate of publishing in this area—indeed, quite the reverse; it seems to have accelerated markedly during the past year or two.

A glance at the index suggests, however, that some topics have received very much more attention than others, and the distribution of entries between the different chronological and subject divisions confirms this. One sees at once, for example, that the work of Maxwell has been subjected to a great deal of scholarly investigation and analysis, as has the question of Newton's influence on 18th-century science. There is likewise a large body of literature on people like Franklin, Faraday, and William Thomson (though the wide disparity between the numbers of index entries for Thomson and Maxwell comes as a surprise). Several other major figures seem seriously under-studied by comparison—Fresnel, Fourier, Helmholtz and Hertz, for example. Similarly, eighteenth-century electricity seems to have attracted a disproportionate amount of attention in comparison with eighteenth-century work on light and heat. Ideas on magnetism and especially sound, from no matter what period, have been investigated even less frequently. Studies of the development, structure, and internal dynamics of communities of physicists in our period and of the exchange and elaboration of information within them have recently started to appear, but the potentialities here have as yet scarcely even begun to be exploited.

Thus despite the recent upsurge in publishing, much remains to be done. Indeed, the works that have appeared have served to open up a wide range of exciting possibilities for future research. This bibliography will, we hope, promote such investigations by providing ready access to the existing literature.

This work has grown out of a project funded by the Australian Research Grants Scheme and would not have been possible without that support. We are likewise grateful for financial assistance provided by the Faculty of Arts, University of Melbourne. Many of the annotations on nineteenth-century heat and thermodynamics have been provided by Dr. Keith Hutchison, whose co-operation and assistance we here gratefully acknowledge. Most of the typing has been done by Lynne Padgham, whose cheerful forbearance in the face of seemingly unending streams of annotated cards has been quite astonishing. To her, too, we offer our thanks.

Anyone who has compiled a bibliography of this kind will know the perennial fear that a major contribution to the field under review will be inadvertently omitted. As Keith Hutchison kindly pointed out to us, the advent of modern word-processing technology provides one with a convenient scapegoat here; all one needs to do is to include an escape clause in one's introduction to the effect that it was discovered at the last moment that the computer had "lost" three entries but that we had been unable to identify which three they were. We thought this an excellent joke until we discovered while compiling the index that Whittaker's famous book (hastily reinserted as item 349a) was missing from our list, even though in this particular case we knew from independent evidence that an entry had been included earlier on! So if any reader feels slighted by not finding his or her work included below, please don't blame us. Blame the machine!

R.W. Home
Mark J. Gittins

University of Melbourne
December 1982

JOURNALS CONSULTED AND ABBREVIATIONS USED

Titles marked with an asterisk were scanned systematically.

Journal	*Abbreviation* (where used)
Acta historiae rerum naturalium necnon technicarum	
*Ambix	
American Historical Review	Amer. Hist. Rev.
American Journal of Physics	Amer. J. Phys.
American Mathematical Monthly	Amer. Math. Monthly
American Physics Teacher	Amer. Phys. Teacher
American Scientist	Amer. Scientist
Annales: économies, sociétés, civilisations	Annales
*Annals of Science	Ann. Sci.
Antiquarian Horology	
Archeion	
*Archive for History of Exact Sciences	Arch. Hist. Exact Sci.
*Archives internationales d'histoire des sciences	Arch. Int. Hist. Sci.
Beiträge zur Geschichte der Technik	
*Berichte zur Wissenschaftsgeschichte	
Bild der Wissenschaft	
*British Journal for the History of Science	Brit. J. Hist. Sci.
British Journal for the Philosophy of Science	Brit. J. Phil. Sci.
*Bulletin of the British Society for the History of Science	Bull. Brit. Soc. Hist. Sci.

Bulletin des Schweizerischen Elektrotechnischen Vereins	*Bull. Schweiz. Elektrotech. Ver.*
Bulletin de la Société Royale des Sciences de Liège	*Bull. Soc. Roy. Sci. Liège*
Bulletin de la Société des Amis d'A.-M. Ampère	*Bull. Soc. Amis A.-M. Ampère*
Bollettino di bibliografia e di storia delle scienze matematiche e fisiche	
Cahiers d'histoire mondiale	*Cah. Hist. Mond.*
*Centaurus	
Contemporary Physics	
Cultura e scuola	
Dix-huitième siècle	
Europa	
*Historia Mathematica	*Hist. Math.*
*Historical Studies in the Physical Sciences	*Hist. Studs. Phys. Sci.*
*History of Science	*Hist. Sci.*
*History of Technology	*Hist. Tech.*
History of Universities	*Hist. Univ.*
*Isis	
Istoriya fiziko-matematicheskikh nauk	
Janus	
*Japanese Studies in the History of Science	*Japanese Studs. Hist. Sci.*
Journal of Interdisciplinary History	*J. Interdisciplinary Hist.*
Journal of the History of Ideas	*J. Hist. Ideas*
Journal of the Institute of Mathematics and its Applications	*J. Inst. Maths. Applics.*
Lychnos	
Medical History	
Memoirs of the Institute of Sciences and Technology, Meiji University	*Mem. Inst. Sci. Tech. Meiji Univ.*
Memoirs and Proceedings of the Manchester Literary and Philosophical Society	*Mem. Proc. Manchester Lit. Phil. Soc.*
Minerva	

Natural Philosopher

NTM. *Schriftenreihe für Geschichte*   NTM
   der Naturwissenschaften,
   Technik und Medizin

Notes and Records of the Royal   Notes Rec. Roy. Soc. Lond.
   Society of London

Organon

Osiris

Oxford Review of Education   Oxford Rev. Educ.

Petrus nonius

Philosophical Journal   Phil. J.

Philosophy of Science   Phil. Sci.

Physis

Proceedings of the American   Proc. Amer. Phil. Soc.
   Philosophical Society

Proceedings of the Huguenot Society   Proc. Huguenot Soc. Lond.
   of London

Rete

Revue de synthèse   Rev. d. synthèse

Revue d'histoire des sciences et de   Rev. Hist. Sci.
   leurs applications

Science

Science Studies

Scientia

Social Studies of Science   Soc. Studs. Sci.

Studies in Eighteenth-Century
   Culture

Studies in History and Philosophy of   Studs. Hist. Phil. Sci.
   Science

Technology and Culture

The Thomist   Thomist

Tijdschrift voor de Geschiedenis der   Tsch. Gesch. Gnk. Natuurw. Wisk.
   Geneeskunde,   Techn.
   Natuurwetenschappen,
   Wiskunde en Techniek

Transactions of the American   Trans. Amer. Phil. Soc.
   Philosophical Society

Verhandelingen der Koninklijk   Verhand. Nederlandsche Akad.
   Nederlandsche Akademie van   Weten.
   Wetenschappen, Amsterdam

The History of Classical Physics

I. GENERAL BIBLIOGRAPHICAL WORKS, BIOGRAPHICAL DICTIONARIES

1. Arago, Dominique François Jean, *Biographies of Distinguished Scientific Men*. Translated by W.H. Smyth, Baden Powell and Robert Grant. 2 vols. London, 1857. Reprinted, Freeport, N.Y.: Books for Libraries Press, 1972.

    Includes an autobiography, together with biographies of Bailly, William Herschel, Laplace (an analysis of his work only, without biographical detail), Fourier, Lazare Carnot, Malus, Fresnel, Young and Watt. Cf. vols. 1-3 of item 111 for the original French versions of most of these essays.

1a. Barr, E. Scott, *An Index to Biographical Fragments in Unspecialized Scientific Journals*. University, Alabama: University of Alabama Press, 1973. vii + 294 pp.

    A listing of biographical notices in the following journals concerning some 7,700 scientists flourishing prior to 1920, presented in alphabetical order with birth and death dates: the *American Journal of Science*, *Nature*, the *Philosophical Magazine*, *Popular Science Monthly*, the *Proceedings* of the Royal Societies of London and Edinburgh, and *Science*.

2. Bierens de Haan, D., *Bibliographie néerlandaise historique-scientifique des ouvrages importants dont les auteurs sont nés aux 16e, 17e et 18e siècles, sur les sciences mathématiques et physiques, avec leurs applications*. Rome: 1883. Reprinted, Nieuwkoop: B. de Graaf, 1960. 424 pp.

    Catalogued by author but also includes a subject index.

3. Bolton, Henry Carrington. *Catalogue of Scientific and Technical Periodicals, 1665-1895, together with Chronological Tables and a Library Check List*. 2nd ed. Washington: 1897. Reprinted, New York: Johnson Reprint, 1965. viii + 1247 pp.

Comprises: an alphabetical listing of the periodicals (including some whose right "to be classed as periodicals is questionable") supplying editor's name, sequence of series and full bibliographical details; chronological tables whereby dates of particular volumes of some 550 of the periodicals may be found; and a subject index. Over eight thousand items are included; as a result, it is more comprehensive than Scudder (item 20), even though it does not include transactions of learned societies.

4. Brush, Stephen G., *Resources for the History of Physics.* Hanover, N.H.: University Press of New England, 1972. x + 86 + vi + 90 pp.

A bibliography with some limited annotations, this work is designed primarily for secondary students and so necessarily has some restrictions on the material it covers. The first section, subdivided into chronological and subject matter categories, lists secondary works. The second section lists "original works of historical importance" providing details of English translations wherever possible. Primarily deals with monographs, but includes some articles and microfilm editions.

5. Debus, Allen G., ed. *World Who's Who in Science: A Biographical Dictionary of Notable Scientists from Antiquity to the Present.* Chicago: Marquis-Who's Who, 1968. xvi + 1855 pp.

Brief but useful entries, giving for each individual date and place of birth and death, family details, and a summary of professional life including important publications and areas of research.

6. Ferguson, Eugene S., *Bibliography of the History of Technology.* Cambridge, Mass.: Society for the History of Technology and M.I.T. Press, 1968. xx + 347 pp.

A valuable reference tool for the historian of physics as well as the historian of technology. Gives a brief but clear description of each work listed.

7. Gillispie, Charles Coulston, editor-in-chief, *Dictionary of Scientific Biography.* 16 vols. New York: Charles Scribner's Sons, 1970-80.

An outstanding item of 20th-century scholarship which has already become an essential research tool for all fields of the history of science, despite some unevenness in the quality of the entries. These include not only the

normal biographical information but also a discussion,
the length of which depends on the editors' assessment of
the individual's importance, of that individual's scientific work. A concise bibliography, usually very valuable
in itself, is appended to each entry. Vol. 15 is a supplementary volume containing, *inter alia*, entries received too late to be included in the main sequence.
These include those for such notables of classical physics
as Laplace and Poisson. Vol. 16 is an index. A one-
volume condensed version has also been published, the
Concise Dictionary of Scientific Biography, ed. James F.
Maurer et al. (New York: Charles Scribner's Sons, 1981).

8. Houghton, Walter E., et al., eds., *The Wellesley Index
to Victorian Periodicals, 1824-1900: Tables of Contents
and Identification of Contributors with Bibliographies
of the Articles and Stories*. 3 vols. Toronto: University of Toronto Press; London: Routledge and Kegan Paul,
1966-79. xxii + 1194 pp; xxii + 1221 pp; xviii +
1012 pp. In progress.

An excellent piece of scholarship that so far has cata-
logued 35 British periodicals. In an overwhelming
majority of cases, an identity is provided for the frequently anonymous authors. As important discussions were
often carried out in these periodicals that were seldom
reprinted in the participants' collected papers, this
can be invaluable.

9. [Institut de France], *Index biographique de l'Académie
de Sciences du 22 décembre 1666 au 1er octobre 1978*.
Paris: Gauthier-Villars, 1979. 513 pp.

Includes a brief historical introduction describing
the different structures the Academy has had since its
inception, followed by a chronological listing from 1795-
1976 of *membres* (subdivided by divisions), *associés
étrangers*, and *correspondants*. This in turn is followed
by an alphabetical listing of all members of any kind
of the Academy with brief biographical notes.

10. Jayawardene, S.A., *Reference Books for the Historian of
Science: A Handlist*. London: Science Museum, 1982.
xiv + 229 pp.

Lists over 1,000 titles including both bibliographical
manuals and general reference books, with comprehensive
author/title and subject indexes.

11. Knight, David, *Sources for the History of Science, 1660-1914*. London: The Sources of History Ltd., 1975. 223 pp.

    A useful introductory survey, though largely limited in scope to British sources.

12. Kronick, David A., *A History of Scientific and Technical Periodicals: The Origins and Development of the Scientific and Technological Press, 1665-1790*. 2nd ed. Metuchen, N.J.: The Scarecrow Press, 1976. xvi + 336 pp.

    Essentially a descriptive listing (considerably augmented since the first edition) of some seven hundred journals that published scientific and medical articles, broken down by the type of journal. Also included are chapters on the historical background of science and on antecedent and similar modern publications. Unfortunately, little analysis is provided of the information set out. Reviewed by N. Reingold, *Isis* 54 (1963): 284-285.

13. Lindroth, Sten, ed., *Swedish Men of Science 1650-1950*. Stockholm: Swedish Institute, Almqvist and Wiksell, 1952.

    Provides biographies of some thirty Swedish scientists including Swedenborg, Wilcke, Ångström, Rydberg, Gullstrand and Siegbahn.

14. MacLeod, R.M., and J.R. Friday, *Archives of British Men of Science*. London: Mansell, 1972. Microfiche.

    Deals mainly with scientists who flourished after 1870.

15. Poggendorff, J.C., *Biographisch-Litterarisches Handwörterbuch zur Geschichte der Exacten Wissenschaften*. 7 Vols. Leipzig: J.A. Barth, 1863-1904; Berlin: Verlag Chemie, 1925-1940; Berlin: Akademie-Verlag, 1955-1962.

    The essential starting point for most investigations in the history of the physical sciences. Each entry gives a biographical summary and list of publications for the scientist concerned.

16. Rechcigl, Miloslav, Jr., "An Introduction to the History of Czechoslovak Thought and Science: A Critical Bibliography." *Isis* 62 (1971): 79-95.

    A bibliographical introduction to materials dealing with, amongst other things, scientific institutions and thought. Lists the standard (Czech) bibliographies, monographs and journals covering these fields.

17. Reuss, Jeremias David, *Repertorium commentationum a societatibus litterariis editarum.* 16 vols. Göttingen, 1801-21. Reprinted, New York: Burt Franklin, 1962.

    An extremely useful early 19th-century classified bibliography of articles appearing in journals published by scientific societies throughout the world to that date. Physics is treated in vol. 4, "mathesis" (including, *inter alia*, mechanics, hydrostatics and hydraulics) in vol. 7.

18. [Royal Society of London], *Catalogue of Scientific Papers, 1800-1900.* 19 vols. Ser. 1-3 (vols. 1-12), London: Eyre and Spottiswoode; John Murray; C.J. Clay and Sons, 1867-1902. Ser. 4 (vols. 13-19), Cambridge: Cambridge University Press, 1914-1925.

    Each series covers a different period. Within each series is an alphabetical list of scientists with details of any publication of theirs that occurred in any of over seventeen hundred periodicals. Also published separately is the *Subject Index* 3 vols. in 4. Cambridge: Cambridge University Press, 1908-14. While incomplete, this fortunately includes the volumes on mathematics (vol. 1), mechanics (vol. 2) and physics (vol. 3, pts. 1 and 2).

19. Russo, François, *Eléments de bibliographie de l'histoire des sciences et des techniques.* 2nd rev. ed. Paris: Hermann, 1969. xvi + 214 pp.

    A valuable general reference, though somewhat biased towards French materials.

20. Scudder, Samuel M., *Catalogue of Scientific Serials of all Countries, including the Transactions of Learned Societies in the Natural, Physical, and Mathematical Sciences, 1633-1876.* Cambridge, Mass., 1879. Reprinted New York: Kraus Reprint Corporation, 1965. xii + 358 pp.

    Lists over four thousand periodicals geographically by country of origin. Provides dates of issue and has a title index of minor subjects. Not as comprehensive as Bolton's similar listing (item 3), but does include transactions of learned societies.

21. Sokolovskaia, Z.K., *200 nauchnykh biografii: Bibliograficheskii spravochnik.* (200 scientific biographies: A

bibliographical reference book.) Moscow: Nauka, 1975.
191 pp.

A survey of the series of biographies of 200 prominent
scientists (about two-thirds of whom are Russian) that
has been published under the auspices of the U.S.S.R.
Academy of Sciences, including full publication details
of each work. The works are directed mainly "toward
advanced students and scholars in the sciences and the
history of science." Although the works listed vary
considerably in quality, many are the only book length
studies on the persons concerned. Reviewed by L.R.
Graham, *Isis* 69 (1978): 434-435.

22. Taylor, E.G.R., *The Mathematical Practitioners of Tudor
and Stuart England*. Cambridge: Cambridge University
Press, for the Institute of Navigation, 1954. xii +
442 pp. *The Mathematical Practitioners of Hanoverian
England, 1714-1840*. Cambridge: Cambridge University
Press, for the Institute of Navigation, 1966. xiv +
502 pp. Bostock, Kate, Susan Hurt and Michael Hurt,
*An Index to the Mathematical Practitioners of Hanoverian
England 1714-1840, by E.G.R. Taylor*. London: Harriet
Wynter, 1980. 23 pp.

Invaluable biographical and bibliographical listings
of approximately three thousand lecturers, writers,
instrument makers, teachers and other contributors to
the mathematical arts. Both of Taylor's books include
useful chronological narrative surveys of the main
branches of the subjects, with cross references to the
(chronologically arranged) biographies. The *Index* of
Bostock, Hurt and Hurt is also very valuable since that
provided in the actual text leaves much to be desired.

23. Thornton, John L., and R.I.J. Tully, *Scientific Books,
Libraries, and Collectors: A Study of Bibliography and
the Book Trade in Relation to Science*. 3rd rev. ed.
London: The Library Association, 1971. x + 508 pp.
Supplement, 1969-1975. London: The Library Association,
1978. viii + 172 pp.

Consists of bibliographical essays containing "the more
important publications" (including both primary and secondary material) on topics such as scientific books of the
18th and 19th centuries, scientific periodicals, and
scientific publishing and bookselling. Includes notes
of reprints and translations, but is necessarily highly
selective in the material included.

24. Wallis, Peter John, *An Index of British Mathematicians: A Checklist. Part 2: 1701-1760*. Newcastle: Project for Historical Biobibliography, University of Newcastle-upon-Tyne, 1976. xxviii + 136 pp.

    "Mathematician" in this context is very broadly defined to include astronomers, instrument makers of all kinds, surveyors, writing masters, etc. Entries give name, dates of birth and death (where known), place(s) of activity, occupation and source(s) of information. Future parts will cover the period prior to 1700 and the period 1761-1850.

25. Warnow, Joan Nelson, *A Selection of Manuscript Collections at American Repositories*. National Catalog of Sources for History of Physics, Report No. 1. New York: American Institute of Physics, 1969. v + 73 pp.

    Useful, although as one would expect there is a strong bias towards 20th-century material.

26. Whitrow, Magda, ed., ISIS *Cumulative Bibliography: A Bibliography of the History of Science formed from* ISIS *Critical Bibliographies 1-90, 1913-1965*. 5 vols. London: Mansell, 1971-82. John Neu, ed., ISIS *Cumulative Bibliography 1966-1975: A Bibliography of the History of Science formed from* ISIS *Critical Bibliographies 91-100 indexing Literature published from 1965 through 1974*. London: Mansell, 1980- . In progress.

    An outstanding scholarly resource. Vols. 1 and 2 of the main series deal with "Personalities" and "Institutions," Vol. 3 with "Subjects," Vols. 4 and 5 with "Civilizations and Periods." Vol. 5 also contains addenda and some corrigenda to Vols. 1-3. The sequel is intended to be completed in 2 volumes, and follows as closely as possible the format and classification scheme devised for the main series. Vol. 1 covers "Personalities and Institutions."

27. Williams, Trevor I., ed., *A Biographical Dictionary of Scientists*. London: Adam and Charles Black, 1969. xii + 592 pp.

    Brief biographies of over one thousand scientists including references for further information.

II. BIOGRAPHIES OF INDIVIDUAL SCIENTISTS

28. Agassi, Joseph, *Faraday as a Natural Philosopher*. Chicago and London: Chicago University Press, 1972. xvi + 359 pp.

An idiosyncratic biography presenting Faraday as a theoretician and methodologist rather than as a pure experimentalist. Sees Faraday as a strikingly independent thinker, while not a revolutionary. Also argues his importance in rejecting the then current Baconian inductivist approach to science. However, the presentation is somewhat discursive, and without prior knowledge of particular aspects of Faraday's work it could be confusing. It is also prone to assess work anachronistically in the light of 20th-century theories, and is inadequately documented. Reviewed by P.M. Heimann, *Isis* 63 (1972): 448-449 and C.A. Ronan, *Ann. Sci.* 32 (1975): 78.

29. d'Albe, E.E. Fournier, *The Life of Sir William Crookes, O.M., F.R.S.* London: 1923/New York: 1924.

The standard life of Crookes (1832-1919) which will remain so due to the loss of the bulk of his manuscripts in 1923. His work embraced spectroscopy, optics and vacuum physics, including investigations on cathode rays and the problems of light pressure and matter theory arising from his invention of the radiometer.

30. Aubry, Paul V., *Monge, le savant ami de Napoléon Bonaparte, 1746-1818*. Paris: Gauthier-Villars, 1934. 380 pp.

An elaborate biography drawing on rich family archives. Takes as its focus Monge's devotion to Napoleon and his cause. Describes his role in the establishment of the École Polytechnique, his work with the Institut and his many administrative positions, and thus complements Taton's *L'oeuvre scientifique de Monge* (item 93).

31. Auerbach, F., *Ernst Abbe: Sein Leben, sein Wirken, seine Personlichkeit*. Leipzig: Akademische Verlagsgesellschaft, 1918. xv + 512 pp.

    Biography of the physicist Abbe (1840-1905) who produced the standard optical theory of image resolution of microscopes while working with the optical firm of Carl Zeiss.

32. Berry, Arthur John, *Henry Cavendish: His Life and Scientific Work*. London: Hutchinson, 1960. 208 pp.

    Most of the work is devoted to a discussion of Cavendish's science, including two chapters on his electrical work and one on heat. The discussion consists largely of slabs of quotation with summarizing connecting paragraphs. No documentation.

33. Blackmore, John T., *Ernst Mach: His Work, Life, and Influence*. Berkeley/Los Angeles: University of California Press, 1972. xx + 414 pp.

    Chiefly concerned with Mach's philosophical work and influence. Also, however, includes a useful account of his more strictly scientific work in psychology and physics.

34. Bowers, Brian, *Sir Charles Wheatstone, F.R.S., 1802-1875*. London: H.M.S.O., 1975. vii + 226 pp.

    An authoritative account of the man and his work, presenting him primarily as an "inventor." Recounts aspects of Wheatstone's friendship with many prominent scientists of his day, including Faraday, and describes his work on acoustics, electricity and telegraphy.

35. Bowley, R.M., *et al.*, *George Green, Miller, Snienton*. Nottingham: Nottingham Castle, 1976. 96 pp.

    Contains useful biographical information in essays on "George Green: His Family and Background" by Frieda M. Wilkins-Jones and "George Green: His Academic Career" by David Phillips, the latter of which is rendered valuable by the inclusion of a number of letters which passed between Green, Sir Edward Ffrench Bromhead and William Whewell. A 7-page introduction on "George Green: His Achievements and Place in Science" is worthless.

36. Brinitzer, Carl, *A Reasonable Rebel: George Christoph Lichtenberg*. Trans. by B. Smith. New York: Macmillan, 1960. 203 pp.

A readable general biography, but with little directly on his scientific work.

37. Broda, Engelbert, *Ludwig Boltzmann: Mensch, Physiker, Philosoph*. Vienna: Franz Deuticke, 1955/Berlin: V.E.B. Deutscher Verlag der Wissenschaften, 1957. viii + 152 pp.

The only monograph on Boltzmann, this is a "semi-popular" work which gives a good account of the man but necessarily omits a detailed analysis of his science. However, it does have some discussion of his work on the foundations of mechanics, Maxwell's theories, and thermo-dynamics, and includes an account of his pragmatic methodology of science.

38. Brown, Sanborn C., *Benjamin Thompson, Count Rumford*. London and Cambridge, Mass.: M.I.T. Press, 1979. xii + 361 pp.

The product of the author's devoted work on Rumford over many years. Reviewed by Brooke Hindle, *Isis* 71 (1980): 512-513.

39. Brunet, Pierre, *Maupertuis: Étude biographique. L'oeuvre et sa place dans la pensée scientifique et philosophique du XVIIIe siècle*. 2 vols. Paris: A. Blanchard, 1929. 199 + 487 pp.

The *Étude biographique* gives a good coverage of Maupertuis' peripatetic life. *L'oeuvre* includes separate sections on his mechanics and his physics (acoustics and optics) as well as a discussion of his epistemology and matter theory.

40. Brunet, Pierre, *La vie et l'oeuvre de Clairaut*. Paris: P.U.F., 1952. vii + 112 pp. (Also published in parts in *Rev. Hist. Sci.* 4 (1951): 13-40, 109-153; 5 (1952): 334-349; 6 (1953): 1-17).

A work completed shortly before the author's death, and published posthumously by his friends and colleagues. No references are provided.

41. Bühler, W.K., *Gauss: A Biographical Study*. Berlin/New York: Springer-Verlag, 1981. viii + 208 pp.

A biography which attempts to link Gauss' life and work with the political, social and technical developments of his times. While not surprisingly emphasizing his mathematics, it also has some account of his physics.

41a. Buttmann, Günther, *The Shadow of the Telescope: A Biography of John Herschel*. Trans. B.E.J. Pagel. New York: Charles

Scribner's Sons, 1970. xiv + 219 pp.

The only substantial biography of the noted astronomer who also made significant contributions to various branches of physics, especially optics, while wielding enormous influence within the British scientific community of his day.

42. Campbell, Lewis, and William Garnett, *The Life of James Clerk Maxwell.* London, 1882. xvi + 662 pp. 2nd ed., abridged and revised, London, 1884. First ed. reprinted with a preface by Robert H. Kargon, New York and London: Johnson Reprint Corp., 1969.

Contains (i) a narrative *Life* by Campbell, a boyhood friend and lifelong correspondent of Maxwell's, which includes many letters to Campbell and others; (ii) an account of Maxwell's scientific work, by Garnett, one of his students at Cambridge; (iii) a selection of his poetry. In the second edition, Garnett's contribution was greatly condensed and distributed throughout the biography, and seven new letters (including four to Faraday) added. These have also been included in the reprint edition.

43. Carnot, Hippolyte, *Mémoires sur Carnot.* 2 vols. Paris, 1861-63.

An indispensable if anecdotal memoir on Lazare Carnot, written by a son.

44. Coulson, Thomas, *Joseph Henry: His Life and Work.* Princeton: Princeton University Press, 1950. viii + 352 pp.

Examines Henry both as a scientist and as an administrator. A somewhat hagiographical account that relies heavily on an unpublished biography of Henry written by his daughter. Reviewed by G.E. Owen, *Isis* 42 (1951): 160-162.

45. Crosland, Maurice, *Gay-Lussac, Scientist and Bourgeois.* Cambridge: Cambridge University Press, 1978. xvi + 333 pp.

An excellent biography that seeks to integrate the social, political and intellectual aspects of Gay-Lussac's career as one of the leaders of French science in the first half of the 19th century. Reviewed by Trevor H. Levere, *Isis* 71 (1980): 298-300 and W.H. Brock, *Ann. Sci.* 37 (1980): 236-237.

46. Dunnington, C. Waldo, *Carl Friedrich Gauss, Titan of Science: A Study of his Life and Work.* New York: Exposition Press, 1955. xvi + 479 pp.

The most extensive biography of Gauss in English including a good coverage of his scientific work.

47. Ellis, George E., *Memoir of Sir Benjamin Thompson, Count
    Rumford, With Notices of his Daughter.* Boston: 1871.
    Reprinted, Boston: Gregg, 1972. x + xvi + 680 pp.

    An invaluable source as it draws on much material that
    has now been lost. However it needs to be used with
    caution as it is rather over-laudatory, does not attempt
    any evaluation of Rumford's science, and for the produc-
    tive Munich years relies primarily on the accounts of
    Rumford and his daughter.

48. Eve, Arthur Stewart and C.H. Creasey, *Life and Work of
    John Tyndall.* London: Macmillan, 1945. xxxiii +
    404 pp.

    The only substantial study on the life of this influ-
    ential and respected contributor to all aspects of 19th-
    century British natural philosophy.

49. Everitt, C.W.F., *James Clerk Maxwell: Physicist and
    Natural Philosopher.* New York: Charles Scribner's
    Sons, 1975. 205 pp.

    An expanded version of the "Maxwell" entry in the
    Dictionary of Scientific Biography. An excellent survey
    of Maxwell's life and scientific work.

50. Forbes, R.J. et al., eds., *Martinus Van Marum: Life and
    Work.* 6 vols. Haarlem: H.D. Tjeenk Willink & Zoon for
    the Hollandsche Maatschappij der Wetenschappen, 1969-
    76.

    A work that is valuable not only for the detailed in-
    formation it provides about Van Marum (1750-1837) but
    also for the broader insights it yields into the style
    of late 18th-century experimental physics. Vol. I in-
    cludes biographical material and a bibliography of Van
    Marum's publications; Vol. II publishes his travel diaries
    from his six journeys to different parts of Europe; Vol.
    III contains essays on various aspects of his scientific
    work; Vol. IV gives a descriptive catalogue of his superb
    collection of scientific instruments and apparatus; Vol.
    V republishes his four most important works, together
    with catalogues of his scientific collections; and
    Vol. VI publishes a selection of his immense scientific
    correspondence.

51. Fuchtbauer, Heinrich von, *Georg Simon Ohm: Ein Forscher
    wachst aus seiner Vater Art.* Berlin: V.D.I. Verlag,
    1939. viii + 246 pp.

The best biography of Ohm. Reviewed by A. Mieli,
Archeion 22 (1940): 341-347.

52. Geike, Archibald, *Memoir of John Michell, M.A., B.D.,*
      F.R.S. Cambridge: Cambridge University Press, 1918.
      108 pp.

     A brief biography of Michell (1724-1793) indicating
     the esteem in which he was held by his scientific con-
     temporaries. Describes his work on artificial magnets
     in which he announced the inverse square law of attrac-
     tion; his work on optics as set out in Priestley's *His-*
     tory ... of Vision, Light, and Colours (including a
     theory of matter similar to Boscovich's); and his ex-
     periment to determine the density of the earth, taken
     from Cavendish's report on his completion of the ex-
     periment after Michell's death.

53. Gibbs, F.W., *Joseph Priestley: Adventurer in Science and*
      Champion of Truth. London: Thomas Nelson, 1965/Garden
      City, N.Y.: Doubleday, 1967. xii + 258 pp.

     One of the best biographies of Priestley, although not
     really adequately documented. Places greatest emphasis
     on his chemical work.

54. Graves, Robert P., *Life of Sir William Rowan Hamilton.*
      3 vols. London: Longmans, Green/Dublin: Hodges, Figgis,
      1882-89.

     A standard Victorian "life," composed mainly of Hamil-
     ton's correspondence from which much of the mathematical
     work and some of the more personal details of Hamilton's
     life have been edited out. Includes a complete bibliog-
     raphy of his work. (cf. item 59).

55. Gray, J.J., and Laura Tilling, "Johann Heinrich Lambert,
      Mathematician and Scientist, 1728-1777." *Hist. Math.*
      5 (1978): 13-41.

     Aims not "to enumerate the high points of Lambert's
     scientific and mathematical work" but to describe it "as
     a unified whole" in the context of the period. Includes
     accounts of his optical and magnetic researches, his
     theory of errors, and, very briefly, his use of graphs
     to analyse his data.

56. Green, H. Gwynedd, "A Biography of George Green, Mathe-
      matical Physicist of Nottingham and Cambridge, 1793-
      1841." *Studies and Essays in History of Science and*

Learning offered in Homage to George Sarton, ed. by
M.F. Ashley Montagu. New York, Henry Schuman, 1946,
pp. 549-594.

A brief (18 pp.) account of Green's career, unfortunately entirely lacking in technical discussion of his work,
together with various documents relating to his life.
Also includes extracts from Kelvin's correspondence concerning his re-discovery of Green's subsequently renowned
*Essay on the Application of Mathematical Analysis to the
Theories of Electricity and Magnetism* (1828).

57. Haas-Lorentz, G.L., ed. *H.A. Lorentz—Impressions of
     His Life and Work.* Amsterdam: North-Holland, 1957.
     172 pp.

Reminiscences of Lorentz by his daughter and others
who knew him, with a chapter on "The Scientific Work,"
without any mathematical details, by A.D. Fokker.

58. Hall, Tord, *Carl Friedrich Gauss.* Translated by Albert
     Froderberg. Cambridge, Mass.: M.I.T. Press, 1970.
     viii + 173 pp.

A general account of Gauss' life, without a detailed
analysis of his scientific work. Useful as one of the
few sources in English, but largely derived from other
secondary sources.

59. Hankins, Thomas L., *Sir William Rowan Hamilton.* Balti-
     more and London: The Johns Hopkins University Press,
     1980. xxii + 474 pp.

A broadly based biography (though not attempting to
supplant Robert Graves' mammoth *Life* [item 54]) which
includes analyses of Hamilton's ray optics, dynamics,
and aether theory as well as his quaternions. Also discusses the influence of his intensive interest in philosophy, especially that of Kant.

60. Henderson, Ebenezer, *Life of James Ferguson, F.R.S., in
     a Brief Autobiographical Account, and Further Extended
     Memoir, with Numerous Notes and Illustrative Engravings
     by Ebenezer Henderson.* Edinburgh: A. Fullarton, 1867,
     503 pp.

This work includes many illustrations of various instruments made by Ferguson (1710-1776), a scientific
lecturer and instrument maker, with descriptions of his
popular books and lectures on all aspects of Newtonian
theoretical and experimental physics.

61. Henderson, James B., *Macquorn Rankine: An Oration*. Glasgow University Publications 26. Glasgow: Jackson, Wylie, 1932. 28 pp.

    One of the few sources for Rankine's life.

62. Herivel, John, *Joseph Fourier: The Man and the Physicist*. Oxford: Clarendon Press, 1975. xii + 350 pp.

    Focuses on the biography and the physics of Fourier—with separate halves of the book devoted to each—deliberately omitting an analysis of his mathematics. The best biographical account of Fourier available, it draws on much unpublished material in a not always successful attempt to place him in a full social, political and scientific context. Includes in an appendix a valuable body of unpublished correspondence. Reviewed by R. Olson, *Isis* 67 (1976): 651-652, I. Grattan-Guinness, *Ann. Sci.* 32 (1975): 503-514 and L. Charbonneau, *Rev. Hist. Sci.* 29 (1976): 63-72.

63. Hildebrandsson, H. Hildebrand, and C.W. Oseen, *Samuel Klingenstiernas Levnad och Werk: Biografisk Skildring*. Stockholm: Almqvist & Wiksell, 1919-25. 88 + 69 pp.

    Part I, "Levnadsteckning," is by Hildebrandsson, and includes the texts of a number of letters by Klingenstierna and others relating to his role in the invention of achromatic lenses. Part II, "Vetenskapliga Arbeten," by Oseen, is concerned entirely with Klingenstierna's mathematical work; an intended third part dealing with his other scientific contributions appears not to have been published.

64. Hindle, Brooke, *David Rittenhouse*. Princeton, N.J.: Princeton University Press, 1964. ix + 394 pp.

    A well-documented biography of the 18th-century American astronomer and instrument maker, examining his scientific and philosophical work and its intellectual context.

65. Jaquel, Roger, *Le savant et philosophe mulhousien Jean-Henri Lambert: Études critiques et documentaires*. Paris: Editions Ophrys, 1977. 170 pp.

    A useful collection of articles previously published in various not always easily accessible journals. None, however, deals directly with Lambert's work on physical subjects.

66. Jolly, W.P., *Sir Oliver Lodge*. London: Constable, 1974.
     256 pp.

     Includes a fairly lengthy study of Lodge's research
     into electromagnetism and ether theory during the period
     1874 to 1900. Reviewed by S. Landefeld, *Isis* 67 (1976):
     321-323.

67. Jones, H. Bence, *Life and Letters of Faraday*. 2 vols.
     London, 1870. x + 427 pp; viii + 499 pp.

     A "biography" constructed almost entirely out of
     Faraday's own letters and journals, with short linking
     passages. The second edition, differently paginated,
     was also published in 1870.

68. Jones, R.V., "Benjamin Franklin." *Notes Rec. Roy. Soc.
     Lond.* 31 (1977): 201-225.

     A brief account of Franklin's life, concentrating on
     his role in Anglo-American relations, mainly in politics
     but with some discussion of his scientific work and cor-
     respondence.

69. Klein, Martin J., *Paul Ehrenfest: Vol. 1, The Making of
     a Theoretical Physicist*. Amsterdam/London: North-
     Holland, 1970. xvi + 330 pp.

     The early chapters describe Ehrenfest's education in
     theoretical physics at the turn of the 20th century,
     chiefly in Vienna under Boltzmann. The projected vol. 2,
     dealing with the latter part of Ehrenfest's career, has
     yet to appear.

70. Knott, C.G., *Life and Scientific Work of Peter Guthrie
     Tait*. Cambridge: Cambridge University Press, 1911.
     x + 379 pp.

     The standard though rather uncritical biographical
     source, this also includes a bibliography of Tait's
     work.

71. Koenigsberger, Leo, *Hermann von Helmholtz*. 3 vols.
     Braunschweig: Friedrich Vieweg und Sohn, 1902-1903.
     (One-vol. English translation and abridgement by
     Frances A. Welby; Oxford: at the Clarendon Press, 1906.
     Reprinted, New York: Dover, 1965.)

     The standard biography, including long extracts from
     Helmholtz' papers and correspondence. (These are con-
     siderably abbreviated in the English edition.)

72. Launay, Louis de, *Le grand Ampère d'après des documents
     inédits*. 2nd ed. Paris: Perrin, 1925. xvi + 278 pp.

     One of the two standard biographies of Ampère. Un-
     fortunately does not devote much detail to his scientific
     work.

73. Leitner, Alfred, "The Life and Work of Joseph Fraunhofer,
     (1787-1826)." *Amer. J. Phys.* 43 (1975): 59-68.

     A concise biography of Fraunhofer including brief ac-
     counts of his work on glass, spectra, and diffraction.
     A good bibliography.

74. Lindeboom, G.A., *Herman Boerhaave: The Man and his Work*.
     London: Methuen/New York: Barnes and Noble, 1968.
     xx + 452 pp.

     As well as being the foremost medical teacher of his
     time, Boerhaave (1668-1737) contributed profoundly to
     the fields of botany and especially chemistry. In this
     last case, his theory of elemental fire was exceedingly
     influential among 18th-century physicists. While this
     work does not treat these aspects of Boerhaave's work in
     any detail, for biographical and medical matters it is
     undoubtedly the definitive biography. Reviewed by R.E.W.
     Maddison, *Ann. Sci.* 25 (1969): 263-264.

75. Livingstone, Dorothy Michelson, *The Master of Light: A
     Biography of Albert A. Michelson*. New York: Charles
     Scribner's Sons, 1973. xi + 376 pp.

     Written by Michelson's youngest daughter, who does not
     attempt to analyse Michelson's scientific achievements.
     Instead, the work provides an excellently researched
     and invaluable portrait of the private man, showing him
     as essentially aloof even with his own family. Also in-
     cludes some valuable unpublished correspondence. Reviewed
     by L.S. Swenson, *Isis* 66 (1975): 432-434.

75a. Lockemann, Georg, *Robert Wilhelm Bunsen: Lebensbild
     eines deutschen Naturforschers*. Stuttgart: Wissenschaft-
     liche Verlagsgesellschaft M.B.H., 1949. 262 pp.

     A straightforward descriptive account of Bunsen's career
     and scientific work.

76. Marković, Zeljko, *Rudje Bošković*. 2 vols. Zabreb: Yugo-
     slav Academy of Arts and Sciences, 1968-69. 1,144 pp.

     The standard life and work, unfortunately available

only in Croatian. Reviewed by E. Stipanić, *Rev. Hist. Sci.* 25 (1972): 69-72.

77. Mendelssohn, K., *The World of Walther Nernst: The Rise and Fall of German Science*. London: Macmillan, 1973. 191 pp.

    Chiefly concerned with the problems of the 20th-century German physics community. Also includes, however, a useful account of Nernst's rise to prominence within the burgeoning German scientific world of the last decades of the classical era.

78. Menshutkin, Boris N., *Russia's Lomonosov: Chemist, Courtier, Physicist, Poet*. Translated by J.E. Thal and E.J. Webster under the direction of W. Chapin Huntington. Princeton, N.J.: Princeton University Press, 1952. x + 208 pp.

    First published in Russian in 1937. In the field of physics Lomonosov (1711-65) is chiefly noted for his theoretical work on matter, electricity and optics. The first two are most extensively discussed here, especially his work on a "kinetic theory of gases" and a mechanical theory of heat. Also discussed are Lomonosov's role in the history of Russian education and his influence on the St. Petersburg Academy of Sciences.

79. Meyer, Kirstine, "The Scientific Life and Work of H.C. Oersted" and "H.C. Oersteds arbejdsliv i det danske samfund" ("H.C. Oersted's Varied Activities in Danish Society"). *H.C. Oersted: Naturvidenskabelige Skrifter*. Ed. K. Meyer (née Bjerrum). Copenhagen: A.F. Hoest and Soen, 1920; vol. 1, pp. xiii-clxvi; vol. 2, pp. xi-clxvi.

    The only comprehensive biography of Oersted. The first part provides an account of Oersted's intellectual background as well as a discussion of his scientific work. The second part (in Danish) is a more general description of his professional life. Summarized by J.R. Nielson, "Hans Christian Oersted--Scientist, Humanist and Teacher." *Amer. Phys. Teacher* 7 (1939): 10-22.

80. Millburn, John R., *Benjamin Martin, Instrument Maker and "Country Showman."* Leiden: Noordhoff International, 1976. xii + 244 pp.

    A good biography of Martin (1705-1781), who was a leading example of the self-educated, itinerant lecturer and

instrument maker not uncommon in England in the 18th century. Includes descriptions of Martin's writings, his
lectures on physics, his instrument-making business, and
his comments on the failings of his rivals.

81. Nordenmark, Nils Viktor Emanuel, *Anders Celsius: Professor
    i Uppsala, 1701-1744.* (In Swedish with French Summary).
    Uppsala: Almqvist & Wiksell, 1936. 284 pp.

    A comprehensive biography drawing on much unpublished
    as well as published material. Provides a good picture
    of the milieu in which Celsius worked, but unfortunately
    does not subject his scientific work to much analysis.

82. Oseen, C.W., *Johan Carl Wilcke, Experimental-Fysiker.*
    Uppsala: Almqvist & Wiksell, 1939. 397 pp.

    An excellent well-documented account of the life and
    work of a leading 18th-century experimental physicist.

83. Peacock, George, *The Life of Thomas Young, M.D., F.R.S.*
    London: 1855.

    Still the definitive source on Young's life, since it
    uses material such as his journal which have since dis-
    appeared.

84. Piacenza, Mario, "Note biografiche e bibliografiche e
    nuovi documenti su G. Beccaria." *Bollettino storico-
    bibliografico subalpino, Turin* 9 (1904): 209-228, 340-
    354.

    In addition to being one of the few biographical
    sources for Beccaria, contains a useful list of his
    unpublished work.

85. Rohr, Moritz von, *Joseph Fraunhofers Leben, Leistungen
    und Wirksamkeit.* Leipzig: Akademische Verlagsgesellschaft,
    1929. xx + 233 pp.

    The standard study. Concentrates largely on Fraun-
    hofer's optical work.

86. Sarton, George, "Lagrange's Personality." *Proc. Amer.
    Phil. Soc.*, 88 (1944): 456-496.

    Draws on a large number of sources to discuss aspects
    of Lagrange's position in the general social and intellec-
    tual climate of his day. Includes a useful bibliography.

86a. Schier, Donald S., *Louis-Bertrand Castel, Anti-Newtonian Scientist*. Cedar Rapids, Iowa: Torch Press, 1941. viii + 229 pp.

Not seen.

86b. Shairp, J.C., P.G. Tait, and A. Adams-Reilly, *Life and Letters of James David Forbes, F.R.S.* London: Macmillan, 1873. xiii + 577 pp.

Not seen.

87. Sharlin, Harold Issadore, in collaboration with Tiby Sharlin, *Lord Kelvin: The Dynamic Victorian*. University Park/London: Pennsylvania State University Press, 1979. xiv + 272 pp.

A biography that aims to explain "the subject's development and significance." The influence of Thomson's father is heavily stressed, and a non-mathematical assessment is given of "those papers which were important in Thomson's intellectual development." Reviewed by Crosbie Smith, *Ann. Sci.* 37 (1980): 687-88 and Elizabeth Garber, *Isis* 71 (1980): 182-183.

88. Snorrason, E., *C.G. Kratzenstein, Professor physices experimentalis Petropol. et Havn. and his Studies on Electricity during the Eighteenth Century*. Odense: Odense University Press, 1974. 206 pp.

A useful study which discusses Kratzenstein's work on electricity and also his activities as a member of the St. Petersburg Academy of Sciences and afterwards as professor of physics at the 18th-century University of Copenhagen. The book is marred, however, by numerous imprecisions and inaccuracies, and must be used with caution.

89. Spiess, Otto, *Leonhard Euler: Ein Beitrag zur Geistesgeschichte des XVIII. Jahrhunderts*. Frauenfeld/Leipzig: Huber, 1929. 228 pp.

A good account of Euler's life, but almost nothing on his scientific work.

90. Spitzer, Paul George, "Joseph John Thomson: An Unfinished Social and Intellectual Biography." Ph.D. thesis, Johns Hopkins University, 1970. 244 pp.

Discusses the social and intellectual milieu in which Thomson rose to prominence, with much detail concerning the style and social position of physics in late Victorian Cambridge, though without reference to Thomson's private papers and correspondence or to other unpublished

material. Analyzes non-mathematically Thomson's publications up to the late 1880s, arguing that they were
very conventional in character.

91. Strutt, Robert John, Fourth Lord Rayleigh, *Life of John
     William Strutt, Third Baron Rayleigh, O.M., F.R.S.*
     London: 1924. Augmented edition with annotations by
     the author and foreword by John N. Howard. Madison
     and London: University of Wisconsin Press, 1968.
     xxvii + 439 pp.

     A good portrayal of Rayleigh as a profound experimen-
     talist showing how he went about his investigations while
     continually drawing upon his earlier work. Includes a
     wealth of personal detail and replaces names deleted in
     the original (1924) edition.

92. Strutt, Robert John, Fourth Lord Rayleigh, *The Life of
     Sir J.J. Thomson, O.M.* Cambridge: at the University
     Press, 1942. x + 299 pp.

     While mostly concerned with "J.J."'s 20th-century
     career, also includes chapters devoted to his early life
     and his first years as Cavendish Professor at Cambridge,
     when he worked chiefly on classical electromagnetic
     theory.

93. Taton, René, *L'oeuvre scientifique de Monge.* Paris:
     Presses Universitaires de France, 1951. 441 pp.

     Though chiefly concerned with Monge's mathematical
     contributions, also provides valuable information on his
     other scientific work and on the institutions of French
     science in the late 18th and early 19th centuries.
     Includes a catalogue of Monge's manuscripts, most of
     which remain in private hands.

94. Taton, René, "Repères pour une biographie intellectuelle
     d'Ampère." *Rev. Hist. Sci.* 31 (1978): 233-248.

     Due to the lack of recent biographical studies on
     Ampère, this is a welcome addition despite its brevity.
     Includes comments on previous biographical works.

95. Thompson, Silvanus P., *The Life of William Thomson,
     Baron Kelvin of Largs.* 2 vols. London, 1910. xx +
     1297 pp.

     A typical "life and letters" in the Victorian style on
     a grand scale, with generous quotations throughout from
     Thomson's letters. Appendices list his various academic
     and other distinctions, his publications (661 items) and
     patents granted to him.

96. Torlais, Jean, *Un physicien au siècle des lumières: l'abbé Nollet, 1700-1770*. Paris: Sipuco, 1954. 271 pp.

   The standard biography of Nollet incorporating the most complete bibliography of his works.

97. Torlais, Jean, *Réaumur: Un esprit encyclopédique en dehors de "l'Encyclopédie."* Rev. ed. Paris: A. Blanchard, 1961. 477 pp.

   The standard biography including, *inter alia*, discussions of Réaumur's work in thermometry and electricity and his place in the Académie Royale des Sciences.

98. Tsverava, G.K., *Georg Wilhelm Richmann (1711-1753)*. Leningrad: "Nauka," 1977. 160 pp. (In Russian.)

   A useful biography of an important but little studied 18th-century experimental physicist.

99. Valson, C.A., *La vie et les travaux de Baron Cauchy*. Paris: 1868. Reprinted 2 vols. in one, with an introduction by R. Taton, Paris: A. Blanchard, 1968. xxxvi + 290; xxiii + 178 pp.

   The standard monograph on Cauchy. Despite the over-laudatory approach, much valuable material is contained in the first-volume "life." The second volume comprises useful summary analyses of Cauchy's publications plus bibliographical details; there are sections on his mechanics and his optics.

100. Van Doren, Carl, *Benjamin Franklin*. London: Putnam/ New York: Viking Press, 1938. xx + 885 pp.

   The standard biography, though written too early to profit from the recent upsurge of scholarly interest in Franklin as a scientist.

101. Wangerin, Albert, *Franz Neumann und sein Wirken als Forscher und Lehrer*. Braunschweig, 1907. x + 185 pp.

   The standard biography.

102. Wheeler, Lynde Phelps, *Josiah Willard Gibbs: The History of a Great Mind*. New Haven: Yale University Press, 1951. xii + 264 pp.

   A useful study including non-technical discussions of different aspects of Gibbs' work.

103. Whyte, Lancelot Law, ed., *Roger Joseph Boscovich, S.J.,*
      F.R.S, 1711-1787: Studies on his Life and Work on the
      250th Anniversary of his Birth. London: Allen and
      Unwin, 1961. 230 pp.

      Included in this collection are articles on the follow-
      ing topics: Boscovichean atomism, by the editor; the
      Theoria, by Seljko Marković; a comparison of Priestley's
      and Boscovich's matter theories, by R.E. Schofield; his
      optics and instrument design, by C.A. Ronan; and a long
      but unfortunately undocumented biographical article, by
      Elizabeth Hill.

104. Wiederkehr, Karl Heinrich, *Wilhelm Eduard Weber: Erforscher*
      der Wellenbewegung und der Elektrizität 1804-91. (Grosse
      Naturforscher, Bd. 32). Stuttgart: Wissenschaftliche
      Verlagsgesellschaft, 1967. 227 pp.

      A straightforward account, without technical detail,
      of Weber's life and scientific work. Based on the
      author's doctoral thesis (item 1187).

105. Williams, L. Pearce, *Michael Faraday: A Biography.* Lon-
      don: Chapman and Hall/New York: Basic Books, 1965.
      xvi + 531 pp.

      An outstanding intellectual biography which carefully
      traces the development, in interaction with his experi-
      mental investigation, of Faraday's theoretical concep-
      tions concerning electricity and magnetism. Some of
      Williams' opinions, especially concerning the influence
      of Boscovich's ideas on Faraday, have proved contro-
      versial, but in general he presents a plausible account
      of a highly original thinker.

106. Winnik, Herbert C., "A Reconsideration of Henry A. Row-
      land--The Man." *Ann. Sci.* 29 (1972): 19-34.

      An attempt to overcome what the author sees as the
      one-sided view of scientists that is general in scientific
      biographies, in this case by investigating Rowland's
      family background and his religious and political views.

107. Wood, Alexander, *Thomas Young, Natural Philosopher,*
      1773-1829. Completed by Frank Oldham. Cambridge:
      Cambridge University Press, 1954. xix + 355 pp.

      A useful supplement to but not a replacement for Pea-
      cock's biography (item 83). The chapters of interest
      for the historian of physics are Ch. 6, "The Professor

of the Royal Institution," Chs. 7 and 8 on the wave
theory of light, and Ch. 11 on Young's *Encyclopaedia
Britannica* articles. While little analysis is given
of the scientific work, there are valuable descriptions
of the multifarious and violent controversies engaged
in by Young. N.H. de V. Heathcote (review, *Isis* 48
(1957): 93-97) warns that there are many silent "corrections" in the quotations given and that these sometimes
change the emphasis of the original passages.

III. COLLECTED WORKS, CORRESPONDENCE AND BIBLIOGRAPHIES OF INDIVIDUAL SCIENTISTS

108. Abbe, Ernst, *Gesammelte Abhandlungen*. 5 vols. Jena: G. Fischer Verlag, 1904-1940.

109. Ampère, André Marie, *Mémoires sur l'électrodynamique*. 2 vols. Paris, 1885-87. (Société français de physique, collection de mémoires relatifs à la physique, vols. 2, 3).

110. [Ampère, André Marie], *Correspondance du grand Ampère*, ed. L. de Launay. 3 vols. Paris: Gauthier-Villars, 1936-43.

    Unfortunately contains many errors of transcription and is not complete. Includes, at the end of volume 2, a complete bibliography of Ampère's works.

111. Arago, Dominique François Jean, *Oeuvres complètes*, ed. J.A. Barral. 17 vols. Paris, 1854-62.

    Cf. item 1.

112. Avogadro, Amedeo, *Opere scelte*. Turin: R. Accademia di Scienze, 1911.

113. [Bernoulli, Jakob], *Jacobi Bernoulli Basiliensis Opera*, ed. Gabriel Cramer. 2 vols. Geneva, 1744. Reprinted, Brussels: Editions Culture et Civilisation, 1967.

114. [Bernoulli, Jakob], *Die Werke von Jakob Bernoulli*. Basel: Birkhäuser, 1969-. In progress.

    Two volumes have so far appeared, namely vol. 1, which deals with Bernoulli's writings in natural philosophy, and vol. 3, which contains his work on probability and statistics.

115. Bernoulli, Johann, *Opera omnia*, ed. Gabriel Cramer. 4 vols. Lausanne and Geneva, 1742. Reprinted, Hildesheim:

Georg Olms Verlag, 1968, with an introduction by J.E. Hofmann.

116. [Bernoulli, Johann], *Der Briefwechsel von Johann Bernoulli*, ed. Otto Spiess. Basel: Birkhäuser, 1955.

Only one volume of this projected multi-volume series has so far appeared, and it appears that, at least in the form envisaged initially, the project has been abandoned.

117. Boltzmann, Ludwig, *Theoretical Physics and Philosophical Problems: Selected Writings*, ed. Brian McGuinness. Dordrecht/Boston: D. Reidel, 1974. xvi + 280 pp. (Vienna Circle Collection, 5).

A collection of Boltzmann's popular writings setting out his conception of the nature of science in general and theoretical physics in particular.

118. Boltzmann, Ludwig, *Wissenschaftliche Abhandlungen*, ed. Fritz Hasenöhrl. 3 vols. Leipzig: J.A. Barth, 1909.

119. [Bouguer, Pierre]. Maheu, Gilles, "Bibliographie de Pierre Bouguer (1698-1758)." *Rev. Hist. Sci.* 19 (1966): 193-205.

Supersedes previous bibliographies.

119a. Bunsen, Robert Wilhelm Eberhard, *Gesammelte Abhandlungen*. 3 vols. Leipzig: Wilhelm Engelmann, 1904.

120. [Carnot, Lazare N.M.], *Oeuvres mathématiques du Citoyen Carnot*. Basel, 1797.

There is no edition of the collected works of Lazare Carnot. This volume contains only his *Essai sur les machines en général* (1783) and the first edition of *Réflexions sur la métaphysique du calcul infinitésimal* (1797), omitting five other published scientific monographs.

121. Cauchy, Augustin Louis, *Oeuvres complètes*. 27 vols. Paris: Gauthiers-Villars, 1882-1958.

122. [Cavendish, Henry], *The Scientific Papers of the Honourable Henry Cavendish*. 2 vols. Vol. 1. *The Electrical Researches*, ed. James Clerk Maxwell, revised by Joseph Larmor. First published Cambridge, 1879. Reprinted, London: Cass, 1979. Vol. 2. *Chemical and Dynamical Works*, ed. T.E. Thorpe. Cambridge: Cambridge University Press, 1921.

123. [Clairaut, Alexis-Claude]. Taton, René, "Inventaire
     chronologique de l'oeuvre d'Alexis-Claude Clairaut
     (1713-1765)." *Rev. Hist. Sci.* 29 (1976): 97-122; "Sup-
     plément à 'l'inventaire de l'oeuvre de Clairaut'
     (1)." Ibid. 31 (1978): 269-271.

     Includes details of Clairaut's publications, a list of
     the principal surviving manuscripts and a provisional
     inventory of his correspondence.

124. Coulomb, Charles Augustin de, *Mémoires*, ed. A. Potier.
     Paris, 1884. (Société français de physique. Collec-
     tion de mémoires relatifs à la physique, vol. 1).

     Unfortunately not a complete collection of Coulomb's
     papers, but does include the main works from the 1780s.

125. Curie, Pierre, *Oeuvres*. Paris: Gauthier-Villars, 1908.

126. Davy, Humphry, *The Collected Works*, ed. John Davy.
     9 vols. London, 1839-40.

127. [Euler, Leonhard]. G. Eneström, "Verzeichnis der
     Schriften Leonhard Eulers." *Jahresbericht Deut. Math.
     Verein., Ergänzungsbande*, Bd. IV. Leipzig: B.G. Teub-
     ner, 1910-13. 388 pp.

     The standard bibliography of Euler's published works.
     An appendix lists the publications of Euler's son, Johann
     Albrecht Euler.

128. [Euler, Leonhard], *Correspondance mathématique et physique
     de quelques célèbres géomètres du XVIIIe siècle; pré-
     cédée d'une notice sur les travaux de Leonard Euler*,
     ed. P.H. Fuss. 2 vols. St. Petersburg, 1843. Re-
     printed, New York and London: Johnson Reprint, 1968.

     Includes correspondence between Euler and Goldbach,
     Johann Bernoulli and Euler, Daniel Bernoulli and Euler,
     and Daniel Bernoulli and Goldbach.

129. [Euler, Leonhard], *Leonhardi Euleri Opera omnia*. Berlin,
     Göttingen, Leipzig, Heidelberg and Zürich: B.G. Teubner,
     1911- . In progress.

     A massive compilation which aims to print in four
     separate series all the works that Euler himself pre-
     pared for publication, his scientific correspondence,
     and his unpublished manuscripts. Ser. 1: *Opera mathe-
     matica*; Ser. 2: *Opera mechanica et astronomica*; Ser. 3:
     Opera physica et miscellanea; Ser. 4: *Commercium epis-
     tolicum et Manuscripta*. Vol. 1 of Ser. 4A is a register

of all Euler's known surviving correspondence, both per-
sonal and scientific, with brief summaries (in German)
of the contents of each letter.

130.   [Euler, Leonhard], *Rukopisnye materialy Leonarda Eilera
       v arkhive akademii nauk U.S.S.R.*, ed. Yu. Kh. Kopele-
       vich et al. 2 vols. Moscow/Leningrad, 1962-65.

       Vol. 1 contains an annotated index of Euler's unpub-
       lished papers in the archives of the Soviet Academy
       of Sciences. Vol. 2 publishes 12 previously unpublished
       papers on mechanics.

131.   [Euler, Leonhard]. G. Eneström, "Der Briefwechsel
       zwischen Leonhard Euler und Johann I Bernoulli."
       Bibliotheca mathematica, Ser. 3, 4 (1903): 344-388;
       5 (1904): 248-291; 6 (1905): 16-87. "Der Briefwechsel
       zwischen Leonhard Euler und Daniel Bernoulli." Ibid.
       7 (1906-7): 126-156.

132.   [Euler, Leonhard], *Eulers und Johann Heinrich Lamberts
       Briefwechsel*, ed. Karl Bopp. Berlin: Akademie der
       Wissenschaften, 1924. (Abhandlungen der Preussischen
       Akademie der Wissenschaften, 1924, 2).

133.   [Euler, Leonhard], *Leonard Eiler: pisma k uchenym*, ed.
       V.I. Smirnov et al. Moscow/Leningrad, 1963.

       Contains letters of Euler to Bailly, Bulfinger, Bonnet,
       Ehler, Wolff and others.

134.   [Euler, Leonhard], *Leonhard Euler und Christian Gold-
       bach: Briefwechsel 1729-1764*, ed. A.P. Yushkevich and
       E. Winter. Berlin: Akademie-Verlag, 1965.

135.   [Euler, Leonhard], *Die Berliner und die Petersburger
       Akademie der Wissenschaften im Briefwechsel Leonhard
       Eulers*, ed. A.P. Yushkevich and E. Winter. 3 vols.
       Berlin: Akademie-Verlag, 1956-76.

135a. [Euler, Leonhard], *The Euler-Mayer Correspondence (1751-
       1755): A New Perspective on Eighteenth-Century Advances
       in the Lunar Theory*, ed. Eric G. Forbes. London:
       Macmillan, 1971. x + 118 pp.

       Though chiefly concerned with astronomical matters,
       these letters also throw useful light on the state of
       the physical sciences more generally in 18th-century
       Germany. The edition includes a useful introduction and

notes by Forbes. The letters have also been published in the original German in *Istoriko-astronomicheskie issledovaniya* 5 (1959): 271-444 and 10 (1969): 285-310.

136. [Faraday, Michael]. Jeffreys, Alan E., *Michael Faraday: A List of his Lectures and Published Writings*. London: Chapman and Hall, for the Royal Institution, 1960. xxviii + 86 pp.

The works are listed chronologically and include his books, articles, papers, letters to the press, lectures and manuscript lecture notes, and reprints. Valuable information on further sources is provided in the introduction.

137. [Faraday, Michael], *Faraday's Diary: being the Various Philosophical Notes of Experimental Investigations made by Michael Faraday*, ed. Thomas Martin. 8 vols. London: G. Bell and Sons, 1932-36.

138. Faraday, Michael, *Experimental Researches in Electricity*. 3 vols. London: B. Quaritch, 1839-55. Reprinted, 3 vols. in 2, New York: Dover, 1965.

139. Faraday, Michael, *Experimental Researches in Chemistry and Physics*. London: Taylor and Francis, 1859. viii + 496 pp.

This and the preceding compilation contain the bulk of Faraday's published research, collected by himself.

140. [Faraday, Michael], *Selected Correspondence*, ed. L.P. Williams. 2 vols. Cambridge: Cambridge University Press, 1971.

141. FitzGerald, George Francis, *Scientific Writings*, ed. with a historical introduction by Joseph Larmor. Dublin: Hodges, Figgis & Co./London: Longmans, Green and Co., 1902.

141a. [Forbes, James David], *An Index to the Correspondence and Papers of James David Forbes (1809-1868) and Also to Some Papers of His Son, George Forbes*. St. Andrews: St. Andrews University Library, 1968. 123 pp.

142. [Fourcault, Jean Bernard Léon], *Recueil des travaux scientifiques de Léon Foucault*, ed. by Mme Foucault and C.M. Gariel. With a "Notice sur les oeuvres de L. Foucault" by J. Bertrand and a "Notice historique" by J.A. Lissajous. 2 vols. Paris: Gauthier Villars, 1878.

Contains Foucault's published and some unpublished work plus the two standard sources for his life and work.

143. [Fourier, Jean-Baptiste-Joseph], *Oeuvres de Fourier*, ed. Gaston Darboux. 2 vols. Paris: Gauthier-Villars, 1888-90.

    The first volume comprises the *Théorie analytique de la chaleur*, the second a number of Fourier's lesser publications.

144. [Franklin, Benjamin]. *The Writings of Benjamin Franklin*, ed. Albert Henry Smyth. 10 vols. New York, 1907. Reprinted, New York: Haskell House Publishers, 1970.

    Long the standard collection of Franklin's writings, has now been largely superseded (cf. item 145 below).

145. [Franklin, Benjamin], *The Papers of Benjamin Franklin*, ed. Leonard W. Labaree et al. New Haven: Yale University Press, 1959-. In progress.

    A superb edition which supercedes all previous collections of Franklin's writings.

146. [Fraunhofer, Joseph von], *Joseph von Fraunhofers gesammelte Schriften*, ed. E.C.J. Lommel. Munich: Königlich Akademie, 1888.

147. Fresnel, Augustin, *Oeuvres complètes*, ed. H. de. Senarmont, E. Verdet and L. Fresnel. 3 vols. Paris: Imprimerie Impériale, 1866-70.

    Includes (vol. 1, pp. ix-xcix) a comprehensive introduction by Emile Verdet and (vol. 3, pp. 475-526) an "Eloge historique" by Arago which is still the most detailed biography of Fresnel.

148. [Galvani, Luigi], *Opere edite ed inedite del professore Luigi Galvani*. Bologna, 1841.

149. Gauss, Carl Friedrich, *Werke*, ed. E.C.J. Scherling et al. 12 vols. Göttingen/Leipzig/Berlin, 1863-1933.

    Vol. 10, pt. 2 and vol. 11, pt. 2 of this collection comprise a number of useful, separately paginated discussions by different authors of various aspects of Gauss' work. These include (vol. 10, pt. 2, sect. 7) Harald Geppert, "Über Gauss' Arbeiten zur Mechanik und Potentialtheorie" (61 pp.), and (vol. 11, pt. 2, sect. 2) Clemens Schaefer, "Über Gauss' physikalische Arbeiten (Magnetismus, Elektrodynamik, Optik)" (217 pp.).

150. Germain, Sophie, *Oeuvres philosophiques ... suivies de pensées et de lettres inédites*. Paris: Paul Ritti, 1879.

151. [Gibbs, Josiah Willard], *The Scientific Papers of J. Willard Gibbs*, ed. H.A. Bumstead and R.G. Van Name. 2 vols. London, 1906. Reprinted, New York: Dover, 1961.

152. Green, George, *Mathematical Papers*, ed. N.M. Ferrers. London: Macmillan, 1871.

153. [Hamilton, William Rowan], *The Mathematical Papers of Sir William Rowan Hamilton*, ed. A.W. Conway et al. 3 vols. Cambridge: Cambridge University Press, 1931-67.

    Vol. 1 contains Hamilton's papers on geometrical optics, Vol. 2 those on dynamics, Vol. 3 those on algebra. A fourth volume is planned.

154. Heaviside, Oliver, *Electrical Papers*. 2 vols. London/New York: Macmillan, 1892. Reprinted, New York: Chelsea Pub. Co., 1970.

155. Heaviside, Oliver, *Electromagnetic Theory*. 3 vols. London: "The Electrician" Printing and Publishing Co., 1894-1912. Reprinted, New York: Chelsea Pub. Co., [1971?].

156. Helmholtz, Hermann Ludwig Ferdinand von, *Wissenschaftliche Abhandlungen*. 3 vols. Leipzig: J.A. Barth, 1882-95.

157. Helmholtz, Hermann Ludwig Ferdinand von, *Vorlesungen über theoretische Physik*, ed. O. Krigar-Menzel, A. König and C. Runge. 6 vols. Leipzig: J.A. Barth/Hamburg: Voss, 1897-1907.

158. [Henry, Joseph], *Scientific Writings of Joseph Henry*. 2 vols. Washington, D.C.: Smithsonian Institution, 1886.

    Long the standard collection of Henry's scientific papers but now gradually being superceded (cf. item 159 below).

159. [Henry, Joseph], *The Papers of Joseph Henry*, ed. Nathan Reingold. Washington, D.C.: Smithsonian Institution Press, 1972-. In progress.

160.   [Henry, Joseph], *A Scientist in American Life: Essays and Lectures of Joseph Henry*, ed. Arthur P. Molella et al. Washington, D.C.: Smithsonian Institution Press, 1980. viii + 136 pp.

       A collection of 13 papers and essays of Henry's on broader questions concerning the nature of science and the place of the scientist within society. Eight have not been published previously.

161.   Herapath, John, *Mathematical Physics and Other Selected Papers*. 2 vols. in 1. Ed. with an introduction by S.G. Brush. New York: Johnson Reprint, 1972.

162.   Herschel, William, *Scientific Papers*. Biographical introduction by J.L.E. Dreyer. 2 vols. London: The Royal Society and the Royal Astronomical Society, 1912.

163.   Hertz, Heinrich Rudolph, *Gesammelte Werke*, ed. Philipp Lenard. 3 vols. Leipzig: J.A. Barth, 1894-95.

       Have been translated into English as: Vol. 1. *Miscellaneous Papers*, trans. D.E. Jones and G.A. Schott. London: Macmillan, 1896; Vol. 2. *Electric Waves*, trans. D.E. Jones. London, 1893, reprinted, New York: Dover, 1962; Vol. 3. *The Principles of Mechanics*, trans. D.E. Jones and T.E. Walley. London, 1899; reprinted New York: Dover, 1956.

164.   Hopkinson, John, *Original Papers*, ed. with memoir by B. Hopkinson. 2 vols. Cambridge: at the University Press, 1901.

165.   [Joule, James Prescott], *The Scientific Papers of James Prescott Joule*. 2 vols. London: The Physical Society of London, 1884-87.

166.   Kirchhoff, Gustav Robert, *Gesammelte Abhandlungen*. Leipzig: J.A. Barth, 1882. *Nachtrag*, ed. Ludwig Boltzmann. Leipzig: J.A. Barth, 1891.

167.   [Lagrange, Joseph Louis]. Taton, René, "Inventaire chronologique de l'oeuvre de Lagrange." *Rev. Hist. Sci.* 27 (1974): 3-36.

       A chronological listing of all the publications of Lagrange including full bibliographical details. An invaluable addition to the faulty *Oeuvres* (item 168).

168. [Lagrange, Joseph Louis], *Oeuvres de Lagrange*, ed. J.A.
      Serret and G. Darboux. 14 vols. Paris: Gauthier-
      Villars, 1867-92. Reprinted, Hildesheim: Georg Olms,
      1973.

169. [Lambert, Johann Heinrich]. Steck, Max, *Bibliographia
      Lambertiana*. 2nd edition, Hildesheim: H.A. Gerstenberg,
      1970. xiv + 123 pp.

      Contains comprehensive primary and secondary bibliog-
      raphies on Lambert and some valuable biographical
      material.

170. [Lambert, Johann Heinrich], *Die handschriftliche Nachlass
      von Johann Heinrich Lambert (1728-1777)*, ed. Max Steck.
      Basel: Öffentliche Bibliothek der Universität, 1977.
      142 pp.

      A catalogue published after Steck's death but based
      on a manuscript of his.

171. Lambert, Johann Heinrich, *Philosophische Schriften*, ed.
      H.W. Arndt. 10 vols. Hildesheim: Georg Olms, 1965.

      The final two volumes are devoted to Lambert's corres-
      pondence.

172. [Laplace, Pierre-Simon], *Oeuvres complètes de Laplace*.
      14 vols. Paris: Gauthier-Villars, 1878-1912.

      Incomplete for published work and omits unpublished
      material and much of the correspondence. In addition
      the mathematical notation has been modernized by the
      editors.

173. [Laplace, Pierre Simon], *Calendar of the Correspondence
      of Pierre Simon Laplace*, ed. Roger Hahn. Berkeley:
      Office for History of Science and Technology, Univer-
      sity of California, 1982. xii + 92 pp.

174. Larmor, Joseph, *Mathematical and Physical Papers*. 2
      vols. Cambridge: at the University Press, 1929.

175. [Lavoisier, Antoine Laurent], Duveen, Denis I., and
      Herbert S. Klickstein, eds., *A Bibliography of the
      Works of Antoine Laurent Lavoisier, 1743-1794*. Lon-
      don: Wm. Dawson & Sons and E. Weil, 1954. xxiv +
      491 pp. Duveen, Denis I., *Supplement to a Bibliog-
      raphy of the Works of Antoine Laurent Lavoisier, 1743-
      1794*. London: Dawsons, 1965. xvi + 173 pp.

176. Lavoisier, Antoine Laurent, *Oeuvres*. 6 vols. Paris: Imprimerie Impériale, 1862-93.

177. [Lavoisier, Antoine Laurent], *Oeuvres de Lavoisier. Tome VII: Correspondance*, ed. René Fric. Paris: Editions Albin Michel, 1955-.

178. [Lavoisier, Antoine Laurent]. Guerlac, Henry, "The Lavoisier Papers—A Checkered History." *Arch. Int. Hist. Sci.* 29 (1979): 95-100.

An account of the dispersal and eventual resting places of various parts of Lavoisier's papers.

179. [Lichtenberg, Georg Christoph]. Jung, Rudolf, *Lichtenberg-Bibliographie*. Heidelberg: Stiehm, 1972. 179 pp.

Very thorough. The majority of the secondary works listed are concerned with Lichtenberg as a literary figure, not as a scientist.

180. [Lichtenberg, Georg Christoph], *Georg Christoph Lichtenbergs vermischte Schriften*, ed. L.C. Lichtenberg and F. Kries. 9 vols. Göttingen, 1800-06. Reprinted, Bern: H. Lang, 1972.

Vols. 6-9 comprise the "Physikalische und Mathematische Schriften."

181. [Lodge, Oliver], Besterman, Theodore, comp., *A Bibliography of Sir Oliver Lodge, F.R.S.* Oxford: Oxford University Press, 1935. xiv + 219 pp.

182. [Lomonosov, Mikhail Vasil'evich], *Polnoe sobranie sochinenii*, ed. S.I. Vavilov. 10 vols. Moscow/ Leningrad: Soviet Academy of Sciences, 1950-59.

183. [Lomonosov, Mikhail Vasil'evich], *Mikhail Vasil'evich Lomonosov on the Corpuscular Theory*. Translated with an introduction by Henry M. Leicester. Cambridge, Mass.: Harvard University Press/London: Oxford University Press, 1970. viii + 280 pp.

A selection of Lomonosov's papers on chemistry and physics relating to the corpuscular theory, showing that he was working in isolation from the ideas and conventions of his contemporaries. Since Lomonosov had no access to a laboratory in the earlier stages of his career, many of these papers appeal to reason rather than to experimental evidence, recalling pre-Newtonian theoretical traditions. A short biography by Leicester is included.

184. Lorentz, Hendrik Antoon, *Collected Papers*, ed. P. Zeeman and A.D. Fokker. 9 vols. The Hague: Martinus Nijhoff, 1934-39.

185. Lorenz, Ludwig Valentin, *Oeuvres scientifiques*, ed. H. Valentiner. 2 vols. Copenhagen: Lehmann and Stage, 1898-1904. Reprinted, New York: Johnson Reprint, 1965.

186. [MacCullagh, James], *The Collected Works of James Mac-Cullagh*, ed. J.H. Jellet and S. Haughton. Dublin: Hodges, Figgis/London: Longmans, Green, 1880.

187. Maupertuis, Pierre Louis Moreau de, *Oeuvres*. 4 vols. Lyon, 1768. Reprinted, Hildesheim: Georg Olms, 1965-74.

188. [Maupertuis, Pierre Louis Moreau de], *Maupertuis et ses correspondants*, ed. Achille Le Seur. Montreuil-sur-mer, 1896. Reprinted, Geneva: Slatkine Reprints, 1971.

Publishes a large number of letters to Maupertuis, and a much smaller number by him. Included are letters from Euler, König, Kästner, Wolff, and Condillac, among others.

189. [Maxwell, James Clerk], *The Scientific Papers of James Clerk Maxwell*, ed. W.D. Niven. 2 vols. Cambridge: Cambridge University Press, 1890. Reprinted, 2 vols. in 1, New York: Dover, 1954.

190. [Maxwell, James Clerk], *Origins of Clerk Maxwell's Electric Ideas, as described in Familiar Letters to William Thomson*, ed. Joseph Larmor. Cambridge: at the University Press, 1937.

191. [Mayer, Julius Robert], *Die Mechanik der Wärme in gesammelten Schriften von Robert Mayer*, ed. J.J. Weyrauch. Stuttgart: J.G. Cotta, 1867.

192. [Mayer, Julius Robert], *Kleinere Schriften und Briefe von Robert Mayer, nebst Mittheilungen aus seinem Leben*, ed. J.J. Weyrauch. Stuttgart: J.G Cottta, 1893.

193. [Mayer, Tobias], *The Unpublished Writings of Tobias
      Mayer*, ed. and trans. with an introduction by Eric G.
      Forbes. 3 vols. Göttingen: Vandenhoeck and Ruprecht,
      1972. (Arbeiten aus der Niedersächsischen Staats-
      und Universitätsbibliothek Göttingen, 9-11.)

      Vol. 3 contains Mayer's work on magnetism.

194. Neumann, Franz Ernst, *Gesammelte Werke*, ed. by his pupils.
      3 vols. Leipzig: B.C. Teubner, 1906-28.

195. Oersted, Hans Christian, *Gesammelte Schriften*. 6 vols.
      Leipzig: C.B. Lorck, 1850-51.

196. Oersted, Hans Christian, *Correspondance ... avec divers
      savants*, ed. H.C. Harding. 2 vols. Copenhagen: H.
      Aschehoug & Co., 1920.

197. Oersted, Hans Christian, *Naturvidenskabelige Skrifter*,
      ed. with two essays on his work by Kirstine Meyer.
      3 vols. Copenhagen: A.F. Hoest and Soen, 1920.

198. [Ohm, Georg Simon], *Gesammelte Abhandlungen von G.S.
      Ohm*, ed. with an introduction by E. Lommel. Leipzig:
      J.A. Barth, 1892.

199. [Playfair, John], *The Works of John Playfair ... with
      a Memoir of the Author*, ed. James G. Playfair. Edin-
      burgh: A. Constable, 1822.

200. Poincaré, Jules Henri, *Oeuvres*, ed. G. Darboux et al.
      11 vols. Paris: Gauthier-Villars, 1916-56.

201. Poynting, John Henry, *Collected Scientific Papers*.
      Cambridge: at the University Press, 1920.

202. [Priestley, Joseph]. Crook, Ronald E., *A Bibliography
      of Joseph Priestley, 1733-1804*. London: Library
      Association, 1966. xiv + 201 pp.

      Section 6, pp. 145-166, lists Priestley's scientific
      writings.

203. [Priestley, Joseph], *Scientific Correspondence of Joseph
      Priestley*, ed. J.C. Bolton. New York, 1892. Reprinted,
      New York: Kraus Reprint, 1969.

      Contains 96 letters of which 65 are not in Schofield's
      edition of the *Scientific Autobiography* (item 204).

204. [Priestley, Joseph], *A Scientific Autobiography of Joseph Priestley, 1733-1804: Selected Scientific Correspondence*, ed. with a commentary by R.E. Schofield. Cambridge, Mass. and London: M.I.T. Press, 1966. xiv + 415 pp.

Consists of selected letters with linking commentary and sections from Priestley's memoirs, showing his widespread communication with other prominent contemporaries, including Franklin, Bergman, Volta and Cavendish.

205. [Rankine, William John Macquorn], *Miscellaneous Scientific Papers by W.J.M. Rankine ... with a Memoir of the Author by P.G. Tait*, ed. W.J. Millar. London: C. Griffin, 1881.

206. [Réaumur, René-Antoine Ferchault de]. Taton, René, "Chronologie de la vie et des oeuvres de René-Antoine Ferchault de Réaumur." *Rev. Hist. Sci.* 11 (1958): 1-12.

207. [Réaumur, René-Antoine Ferchault de]. Torlais, Jean, "Inventaire de la correspondance et des papiers de Réaumur conservés aux Archives de l'Académie des Sciences de Paris." *Rev. Hist. Sci.* 12 (1959): 315-326.

A valuable guide to Réaumur's papers, which are, however, mainly concerned with biological matters.

208. Reynolds, Osborne, *Papers on Mathematical and Physical Subjects*. 2 vols. Cambridge: Cambridge University Press, 1900-1901.

209. Riemann, Georg Friedrich Bernhard, *Gesammelte mathematische Werke und wissenschaftlicher Nachlass*, ed. R. Dedekind and H. Weber. Leipzig: B.G. Teubner, 1876.

210. [Righi, Augusto], *Scelta di scritti di Augusto Righi*, ed. G.C. Dalla Noce and G. Valle. xxxi + 374 pp. Bologna: Zanichelli, 1950.

211. Ritter, Johann Wilhelm, *Physisch-Chemische Abhandlungen in chronologischer Folge*. 3 vols. Leipzig: C.H. Reclam, 1806.

212. [Ritter, Johann Wilhelm], *Die Begründung der Elektrochemie und Entdeckung der ultravioletten Strahlen:*

Eine Auswahl aus den Schriften des romantischen Physikers, ed. Armin Hermann. Frankfurt am Main: Akademische Verlagsgesellschaft, 1968. 105 pp. (Ostwalds Klassiker der exakten Wissenschaften, N.S., vol. 2).

Gives extracts from Ritter's most important scientific papers, with valuable editorial introduction and commentary.

213. [Ritter, Johann Wilhelm], *Briefe eines romantischen Physikers: Johann Wilhelm Ritter an Gotthilf Heinrich Schubert und an Karl von Hardenberg*, ed. Friedrich Klemm and Armin Hermann. Munich: Heinz Moos Verlag, 1966. 67 pp.

Eight previously unpublished letters of Ritter's, with valuable editorial commentary.

214. Ritz, Walther, *Gesammelte Werke/Oeuvres*. Paris: Gauthier-Villars, 1911.

215. [Rowland, Henry Augustus], *The Physical Papers of Henry Augustus Rowland ..., 1876-1901*. Baltimore: Johns Hopkins University Press, 1902.

216. Sommerfeld, Arnold J.W., *Gesammelte Schriften*, ed. F. Sauter. 4 vols. Braunschweig: F. Vieweg, 1968.

217. Stokes, George Gabriel, *Mathematical and Physical Papers*, ed. George G. Stokes and Joseph Larmor. 5 vols. Cambridge, 1880-1905. Reprinted with a preface by C.A. Truesdell, London and New York: Johnson Reprint, 1966.

218. [Stokes, George Gabriel], *Memoir and Scientific Correspondence of the Late Sir George Gabriel Stokes*, ed. Joseph Larmor. 2 vols. Cambridge, 1907. Reprinted, London and New York: Johnson Reprint, 1971.

219. Strutt, John William, Third Lord Rayleigh, *Scientific Papers*. 6 vols. Cambridge, 1899-1920. Reprinted, New York: Dover, 1964.

220. Tait, Peter Guthrie, *Scientific Papers*. 2 vols. Cambridge: Cambridge University Press, 1898-1900.

221. Thompson, Benjamin, Count Rumford, *The Complete Works of Count Rumford*. 4 vols. Boston: American Academy of Arts and Sciences, 1870-75.

222. [Thompson, Benjamin, Count Rumford], *The Collected Works of Count Rumford*, ed. Sanborn C. Brown. 5 vols. Cambridge, Mass.: Belknap Press/London: Oxford University Press, 1968-70.

A new edition with the papers collected by subject rather than chronologically as in the earlier edition. However, the edition is not comprehensive; see R.J. Morris' review, *Isis* 62 (1971): 412-413.

223. Thomson, William, 1st Baron Kelvin of Largs, *Reprint of Papers on Electrostatics and Magnetism*. London: Macmillan, 1872.

224. Thomson, William, 1st Baron Kelvin of Largs, *Mathematical and Physical Papers*, ed. W. Thomson and J. Larmor. 6 vols. Cambridge: at the University Press, 1882-1911.

224a. [Thomson, William, 1st Baron Kelvin of Largs], *Kelvin Papers: Index to the Manuscript Collection of William Thomson, Baron Kelvin in Glasgow University Library*. Glasgow: Glasgow University Library, 1977. 93 pp.

224b. [Verdet, Émile], *Oeuvres d'Émile Verdet*. 8 vols. Paris: Imprimerie Nationale, 1868-72.

Vol. 1 includes a "Notice sur Émile Verdet" by A. de la Rive.

225. [Volta, Alessandro], *Le opere di Alessandro Volta*. Edizione Nazionale. 7 vols. Milan: Hoepli, 1918-29. Reprinted, New York: Johnson Reprint, 1968.

226. Volta, Alessandro, *Epistolario*. Edizione Nazionale. 5 vols. Bologna: Zanichelli, 1949-55. *Aggiunte alle opere e all'epistolario di Alessandro Volta*. Edizione Nazionale. Bologna: Zanichelli, 1966. *Indici delle opere e dell'epistolario*. Edizione Nazionale. 2 vols. Milan: Rusconi, 1974-76.

227. [Volta, Alessandro], *Opere scelte di Alessandro Volta*, ed. Mario Gliozzi. Turin: Tipografia torinese, 1967.

Includes (pp. 9-30) a very useful overview by Gliozzi of Volta's scientific work.

228. [Waterston, John James], *The Collected Scientific Papers of John James Waterston*, ed. with a biography by J.S. Haldane. London and Edinburgh: Oliver and Boyd, 1928. lxviii + 709 pp.

229. Weber, Wilhelm Eduard, *Werke*. 6 vols. Berlin: J. Springer,
      1892-94.

230. [Wheatstone, Charles], *The Scientific Papers of Sir
      Charles Wheatstone*. Published by the Physical Society
      of London. London: Taylor and Francis, 1879.

231. Young, Thomas, *Miscellaneous Works of the Late Thomas
      Young, M.D., F.R.S.*, ed. G. Peacock and J. Leitch.
      3 vols. London, 1855. Reprinted, New York: Johnson
      Reprint, 1972.

IV. GENERAL HISTORIES

(a) General

232. d'Abro, A., *The Rise of the New Physics, its Mathe-matical and Physical Theories*. 2nd ed. (First edi-tion titled *The Decline of Mechanism*.) 2 vols. New York: 1939. Reprinted, New York: Dover, 1951.

This is not a history but rather an account of differ-ing methodologies and concepts that have evolved in physics at various times, including a discussion of the developments in mathematics that have been involved in 19th and 20th-century physics. Vol. 1 is concerned with classical physics.

233. Ball, W.W. Rouse, *A History of the Study of Mathematics at Cambridge*. Cambridge: Cambridge University Press, 1889. xvi + 264 pp.

Describes the achievements of various Cambridge mathe-maticians and provides some account of the courses they provided and texts they used up to the mid-19th century. Includes separate chapters describing the educational and assessment systems, the subjects in the mathematics tripos from *c.* 1725, and a brief institutional and social history of Cambridge.

234. Bellone, Enrico, *A World on Paper: Studies on the Second Scientific Revolution*. Cambridge, Mass.: MIT Press, 1980. 220 pp.

Published initially in Italian as *Il mondo di carta* (1976). A collection of five essays dealing with dif-ferent phases of classical physics in its heyday, and concerned especially with the relationship between mathe-matics and experience in physical theory. Kelvin's attitude to theory-construction, and the criticism that he and P.G. Tait levelled against Boltzmann's approach, provide the leit-motif. The work ranges widely, however, and also includes discussions on 18th and early 19th-century views such as those of Nollet, Coulomb and Ampère.

235. Blake, R.M., et al., eds., *Theories of Scientific Method: The Renaissance Through the 19th Century*. Seattle: University of Washington Press, 1960. vi + 346 pp.

Includes chapters describing the philosophy of science of John Herschel, William Whewell and John Stuart Mill.

236. Bochner, Salomon, *The Role of Mathematics in the Rise of Science*. Princeton, N.J.: Princeton University Press, 1966. x + 386 pp.

A series of essays, some concerned with the necessity of concepts in mathematics such as those of function, complex numbers and products, for the development of concepts in physics.

237. Cajori, Florian, *A History of Physics in its Elementary Branches Including the Evolution of Physical Laboratories*. London/New York: Macmillan, 1899. viii + 322 pp. Revised ed., 1929. 438 pp. Reprinted, New York: Dover, 1962.

Despite its relative age, this is still a worthwhile, readable but not comprehensive introduction to the history of physics (the treatment of mechanics is at best inadequate). Nearly two thirds of the work concerns the 18th and 19th centuries, subdivided into the basic subject areas of heat, sound, light, and electricity and magnetism.

238. Cantor, G.N., and M.J.S. Hodge, eds., *Conceptions of Ether: Studies in the History of Ether Theories, 1740-1900*. Cambridge: Cambridge University Press, 1981. x + 351 pp.

Contains a 60 pp. introduction by the editors on "Major themes in the development of ether theories from the ancients to 1900," together with 10 previously un-published essays as follows: P.M. Heimann, "Ether and imponderables"; J.R.R. Christie, "Ether and the science of chemistry, 1740-1790"; Roger K. French, "Ether and physiology"; G.N. Cantor, "The theoretical significance of ethers"; Larry Laudan, "The medium and its message: a study of some philosophical controversies about ether";

J.L. Heilbron, "The electrical field before Faraday";
Jed Z. Buchwald, "The quantitative ether in the first
half of the nineteenth century"; Daniel M. Siegel,
"Thomson, Maxwell, and the universal ether in Victorian
physics"; M. Norton Wise, "German concepts of force,
energy, and the electromagnetic ether, 1845-1880";
Howard Stein, "'Subtler forms of matter' in the period
following Maxwell."

239. Cardwell, D.S.L., *Technology, Science and History: A
Short Study of the Major Developments in the History
of Western Mechanical Technology and their Relationships with Science and Other Forms of Knowledge.*
London: Heinemann, 1972. xii + 244 pp.

Arranged, apart from chapters on the medieval background, Galileo's mechanical laws, and 20th-century
events, around the theme of the evolution of the heat
engine in terms of theories and their practical applications. Is thus more restricted in scope than the title
indicates, and does not take into account many broader
scientific, technical and social influences. Reviewed
by R.A. Buchanan, *Brit. J. Hist. Sci.* 6 (1972-73): 314-
315.

240. Clagett, Marshall, ed., *Critical Problems in the History
of Science: Proceedings of the Institute for the
History of Science at the University of Wisconsin,
September 1-11, 1957.* Madison: University of Wisconsin Press, 1959. xiv + 555 pp.

Relevant papers from this noteworthy collection are
detailed separately below.

241. Crosland, Maurice P., ed., *The Science of Matter, a
Historical Survey: Selected Readings.* Harmondsworth:
Penguin Education, 1971. 440 pp.

Comprises 163 brief readings from original sources,
with full references, arranged in roughly chronological
order. The 18th-century section concerns mainly chemical
models and the 19th-century section, while better
balanced, unfortunately omits the important British
aether theories.

242. Crosland, Maurice P., ed., *The Emergence of Science in
Western Europe.* London: Macmillan, 1975. 201 pp.

A collection of papers examining aspects of science
in the 17th and 18th centuries in various national con-

texts, looking at its social, intellectual and institutional background. The papers are listed separately below.

243. Gerland, E., and F. Traumüller, *Geschichte der physikalischen Experimentierkunst*. Leipzig, 1899. Reprinted, Hildesheim: Georg Olms, 1965. xvi + 442 pp.

   While there is an emphasis on 17th century work, almost a third of this useful book is devoted to 18th and 19th-century material, concentrating on Germanic contributions. Includes over 400 illustrations.

244. Gillispie, Charles Coulston, *The Edge of Objectivity: An Essay in the History of Scientific Ideas*. Princeton, N.J.: Princeton University Press, 1960. 562 pp.

   A justly esteemed work which includes chapters on "Science and the Enlightenment" (pp. 151-201), "Early Energetics" (pp. 352-405) and "Field Physics" (pp. 406-492). The final chapter (pp. 493-520) discusses *fin-de-siècle* philosophy of physics, culminating in the work of Einstein.

245. Grigor'ian, A.T., and A.P. Yushkevich, with T.N. Klado and Yu. Kh. Kopelevich, eds., *Russko-frantsuzskie nauchnye svyazi*. Leningrad: Nauka, 1968. 298 pp.

   Publishes a variety of documents from French and Russian archives illustrating scientific links between the two countries since 1717, namely documents on the elections of Russian scientists to the Paris Academy of Sciences and of French scientists to the Russian Academy, and the extensive correspondence between L. Euler and the astronomer J.N. Delisle, in the original French and in Russian translation. Physicists concerning whom documents are published include Poncelet, Poincaré, Langevin and Perrin.

246. Hall, A. Rupert, *The Scientific Revolution, 1500-1800. The Formation of the Modern Scientific Attitude*. London, New York and Toronto: Longmans, Green, 1954. xvii + 390 pp.

   The standard introductory account of the period.

247. Harig, Gerhard, ed., *Sowjetische Beiträge zur Geschichte der Naturwissenschaft*. Berlin: VEB Deutscher Verlag der Wissenschaften, 1960. viii + 243 pp.

A collection of eight previously unpublished essays
by workers from the Institute of the History of Science
and Technology of the Soviet Academy of Sciences.
Papers relating to classical physics are noted individually elsewhere in this bibliography.

248. Hesse, Mary B., *Forces and Fields: The Concept of Action
at a Distance in the History of Physics*. London:
Thomas Nelson & Sons, 1961. x + 318 pp.

Traces various answers that have been given from
primitive times to the 20th century to the question:
"How do bodies act on one another across space?" Chaps.
VII and VIII, on "Action at a Distance" and "The Field
Theories" respectively (pp. 157-225) deal with 18th and
19th-century views.

248a. Hooykaas, Reijer, "Von der 'Physica' zur Physik."
Humanismus und Naturwissenschaften, ed. R. Schmitz
and F. Krafft. Boppard am Rhein: Boldt, 1980, pp.
9-38.

Not seen.

249. Jammer, Max, *Concepts of Force: A Study in the Foundations of Dynamics*. Cambridge, Mass.: Harvard University Press, 1957. x + 269 pp.

Includes chapters on dynamical and theological explanations of force, attempts to explain the nature of gravitational force, and philosophical criticisms of force
in the 18th and 19th centuries.

250. Kangro, Hans, *Geschichte der Physik: Renaissance bis
zum 18. Jahrhundert*. Lüneburg: Nordland-Druck GmbH,
1978. 117 pp.

A posthumously edited version of the author's course
delivered at Hamburg University. Presents a rapid
survey of the main developments during the period in
the various branches of physics as presently defined.

250a. Krafft, Fritz, "Der Weg von den Physiken zur Physik an
den deutschen Universitäten." *Berichte zur Wissen-
schaftsgeschichte* 1 (1978): 123-162.

Surveys with much valuable detail the changing conception of 'physics' within the German universities
from the 17th to the mid-19th century.

251.   Kuhn, Thomas S., *The Structure of Scientific Revolutions*.
       Chicago, 1962. 2nd, enlarged ed., Chicago: University
       of Chicago Press, 1970. xii + 210 pp.

       A justly famous work that has given rise ever since
       its first publication to fierce debates among both
       historians and philosophers of science, and that has
       inspired, either directly or indirectly, much of the
       renewed interest of historians of science in recent
       years in the social organization as well as the intel-
       lectual content of the science of the past. Portrays
       most scientific work as puzzle-solving within a generally
       accepted intellectual framework, or "paradigm," this
       being interrupted occasionally by a "scientific revolu-
       tion" in which one paradigm is replaced by another.
       Many of the examples upon which the argument is based
       are drawn from the history of 18th and 19th-century
       physics.

252.   Kuhn, Thomas S., "Mathematical versus Experimental
       Traditions in the Development of Physical Science."
       J. Interdisciplinary Hist. 7 (1976): 1-31. Reprinted
       in *The Essential Tension: Selected Studies in Scien-
       tific Tradition and Change* by Thomas S. Kuhn (Chicago/
       London: University of Chicago Press, 1977), pp. 31-65.
       First published in French translation, in *Annales* 30
       (1975): 975-998.

       Argues that there was traditionally a sharp separation
       within the physical sciences between the "classical,"
       mathematically based areas of investigation--astronomy,
       statics (including hydrostatics), music, optics and,
       eventually, mechanics--and the experimentally based
       "Baconian" fields such as heat, electricity and magnetism,
       which first became distinct foci for scientific activity
       in the 17th century. Suggests that these two traditions
       remained largely separate throughout the 18th century,
       eventually becoming united in the early 19th century
       into a new discipline, physics.

253.   Kuhn, Thomas S., *The Essential Tension: Selected Studies
       in Scientific Tradition and Change*. Chicago/London:
       University of Chicago Press, 1977. xxiv + 366 pp.

       An important collection of papers, some previously
       unpublished, containing numerous invaluable insights
       into the history of the physical sciences. Several are
       detailed separately below. Essay review by Ian Hacking,
       History and Theory 18 (1979): 223-236.

254. Laudan, Laurens, "Theories of Scientific Method from
        Plato to Mach: A Bibliographical Review." *Hist. Sci.*
        7 (1968): 1-63.

        Pp. 24-38 briefly describe the main methodologies of
        the 18th and 19th centuries, showing that most 18th-
        century natural philosophers strongly advocated "Newtonian-
        Baconian" inductivism, this being later ameliorated in
        Britain and France by tendencies toward Kantian idealism.
        A large bibliography, pp. 38-63.

255. Laudan, Larry, *Science and Hypothesis: Historical Essays
        on Scientific Methodology*. Dordrecht/Boston/London:
        D. Reidel, 1981. xii + 258 pp.

        A collection of historical essays, mostly previously
        published elsewhere, on questions of scientific method.
        A paper on "The Epistemology of Light: Some Methodological
        Issues in the Subtle Fluids Debate" (pp. 111-140)
        describes various 18th-century debates on the admissi-
        bility of theories invoking an aether, before arguing
        that some major 19th-century thinkers were induced to
        develop a new attitude towards scientific hypotheses
        by the manifold successes of the wave theory of light.
        Another paper discusses (pp. 202-225) "Ernst Mach's
        Opposition to Atomism."

256. Magie, W.F., *A Source Book in Physics*. New York/London:
        McGraw-Hill, 1935. xiv + 620 pp.

        Gives "the most significant passages" from the works
        of leading contributors since the time of Galileo to
        the science of physics (in the modern sense of the word).
        Entries are preceded by brief, chiefly biographical notes
        and are arranged by subject--"mechanics," "properties
        of matter," "sound," and so on.

257. Mason, S.F., *A History of the Sciences: Main Currents
        of Scientific Thought*. New York: Henry Schuman/
        London: Routledge and Kegan Paul, 1953. vii + 520 pp.

        One of the better general histories of science. Parts
        IV and V deal respectively with 18th and 19th-century
        science.

258. Olson, Richard, "Scottish Philosophy and Mathematics,
        1750-1830." *J. Hist. Ideas* 32 (1971): 29-44.

        Argues that the use in Scotland of geometrical rather
        than analytical mathematics, despite the acknowledged
        virtues of the latter, was due to the domination of

Common Sense philosophy, which argued for an ultimately physical--and hence geometrical--source of mathematical
ideas, and which saw mathematics as providing training
in awareness of the processes of reasoning, this being
allowed more easily by geometrical than analytical
methods.

259.   Rainoff, T.J., "Wave-Like Fluctuations of Creative Pro-
       ductivity in the Development of West-European Physics
       in the Eighteenth and Nineteenth Centuries." *Isis* 12
       (1929): 287-319.

       Graphical analysis of the number of scientific "dis-
       coveries" (drawn from secondary sources) made in
       Britain, France and Germany during the period stated.
       Claims that these occur in "waves" which seem to be
       linked to similar economic "waves."

260.   Rosenberger, Ferdinand, *Die Geschichte der Physik*.
       3 vols. Braunschweig, 1882-90. Reprinted, 3 vols.
       in 2, Hildesheim: Georg Olms, 1965.

       Still a valuable source, especially for German
       physics. Vol. 2 covers the period 1600 to c. 1780,
       vol. 3, 1780 to c. 1880.

261.   Shapin, Steven, and Arnold Thackray, "Prosopography as
       a Research Tool in History of Science: The British
       Scientific Community 1700-1900." *Hist. Sci.* 12 (1974):
       1-28.

       The authors argue that there is an inherent anachronism
       in standard methods of writing history of science and
       advocate prosopography--"employing collective biography
       ... [as] a sophisticated tool for establishing links
       between action and context"--as an alternative. They
       provide a wide-ranging bibliographical sketch of mainly
       biographical sources, 1700-1900.

262.   Stein, Howard, "On the Notion of Field in Newton, Max-
       well and Beyond," pp. 264-287 in Roger H. Stuewer, ed.,
       Historical and Philosophical Perspectives of Science.
       (Minnesota Studies in the Philosophy of Science, 5).
       Minneapolis: University of Minnesota Press, 1970.

       Mainly concerned with the metaphysical concepts in-
       herent in Newton's "field theory," but includes a short
       discussion of the partial breakdown of those concepts
       in the development of Maxwell's theories. The paper is
       followed by comments by Gerd Buchdahl and Mary Hesse,
       and a reply by Stein.

263. Snelders, H.A.M., "Physics and Chemistry in the Netherlands in the Period 1750-1850." *Janus* 65 (1978): 1-20.

    A useful survey which concentrates especially on (i) the work of J.H. Van Swinden (1746-1823) on electricity and magnetism and (ii) the reception of electromagnetism and electrodynamics in Holland.

264. Taton, René, ed., *A General History of the Sciences*. 4 vols. trans. A.J. Pomerans. London: Thames and Hudson, 1965.

    Published initially in French as *Histoire générale des sciences* (4 vols., 1957-64). A useful but limited series of historical essays on different subjects by a variety of authors attempting to cover the whole sweep of the history of science. Vol. 2, *The Beginning of Modern Science 1450-1800*, and Vol. 3, *The Nineteenth Century* are relevant to the period of this bibliography.

265. Tilling, Laura, "Early Experimental Graphs." *Brit. J. Hist. Sci.* 8 (1975): 193-213.

    Describes some 18th and early 19th-century uses of graphs to plot experimental and observational data. Argues that until the work of J.D. Forbes in the 1830s, graphs were seldom used to analyze data. On the contrary, "usually they arose unintentionally out of the method of measurement used." J.H. Lambert's work is an exception in this regard, and is discussed in some detail.

266. Tilling, Laura, "The Interpretation of Observational Errors in the Eighteenth and Early Nineteenth Centuries." Ph.D. thesis, University of London (Imperial College), 1973. 420 pp.

    Argues that during the 18th century there were "three quite separate threads of development" of relevance to the history of error theory, namely (i) an insistence on the primacy of measurement, based, however, on a "very naive" methodology devoid of a theory of errors; (ii) a number of simple algorithms developed for treating errors in a few quite specific cases; and (iii) a mathematical theory of errors, divorced however from practical measurement systems. Describes a first attempt, by J.H. Lambert in the 1760s, at tying these separate strands together, before concluding with an extended discussion of the work of Gauss and Laplace early in the following century.

267. Vucinich, Alexander, *Science in Russian Culture. Vol.*
      I: A History to 1860; Vol. II: 1861-1917. Stanford,
      Calif.: Stanford University Press, 1963-70. xvi +
      463 pp.; xvi + 575 pp.

      An excellent general survey.

268. Walter, Emil J., *Soziale Grundlagen der Entwicklung*
      der Naturwissenschaften in der älten Schweiz. Bern:
      Francke, 1958. 383 pp.

      A detailed account of the social setting of Swiss
      science from the Reformation to the end of the 18th
      century, including, *inter alia*, much information on
      the social background of major contributors to 18th-
      century physics such as the members of the Basle school.

                   (b) Institutional Histories

269. Bertrand, Joseph, *L'Académie des Sciences et les acadé-*
      miciens de 1666 à 1793. Paris: 1869. Reprinted,
      with index, Amsterdam: B.M. Israel, 1969. 435 + [8]
      pp.

      One of the three "standard" histories of the Academy,
      this concentrates on the working of the society and
      the relationship of this to the activities of various
      of its members.

270. Crosland, Maurice, "History of Science in a National
      Context." *Brit. J. Hist. Sci.* 10 (1977): 95-113.

      An appeal for the study of science in its national
      context. Briefly examines 19th-century France as an
      example, isolating patterns of education, institutionali-
      zation, centralization, professionalization, specializa-
      tion and governmental support as factors that need to
      be taken into account.

271. Gunther, R.T., *Early Science in Cambridge.* Oxford:
      Oxford University Press, 1937. xii + 513 pp.

      Includes a fifty-page chapter giving descriptions and
      illustrations of physical apparatus used at Cambridge
      up till the end of the 19th century, supplemented by
      brief accounts of various researches carried out at the
      university.

272. Gunther, R.T., *Early Science in Oxford*. 14 vols. Oxford: Oxford University Press, 1920-45.

      Vol. 1 contains some account of discoveries in physics in the 18th and 19th centuries, an account of 18th century teaching of physics and descriptions of various pieces of apparatus at the university. Also includes similar accounts of mathematical instruments and a list of instrument makers.

273. Harnack, Adolf von, *Geschichte der königlich preussischen Akademie der Wissenschaften zu Berlin*. 3 vols. in 4. Berlin, 1900.

      Vol. 1, pts. 1 and 2, is a systematic history of the Academy during its first 200 years, the emphasis being on institutional matters. Vol. 2 publishes, as supporting evidence, a large body of documentary material from the Academy's archives and elsewhere. Vol. 3 is a catalogue, arranged both by author and by subject matter, of memoirs and addresses published in the Academy's journals, 1700-1899.

274. Inkster, Ian, "A Note on Itinerant Science Lecturers, 1790-1850." *Ann. Sci.* 28 (1972): 235-236.

      A preliminary report on the author's investigation of the itinerant lecturer tradition in Britain. Claims that itinerant lecturers were far more influential, both in terms of their subject matter and in geographical coverage, than previously believed.

275. Kjoelsen, Hans H., *Fra Skidenstraede til H.C. Oersted Institutet*. Copenhagen: Gjellerups Forlag, 1965. 247 pp.

      Contains a wealth of accurately documented, not usually accessible, information tracing the development of laboratories for physics and chemistry in higher education institutions in Copenhagen from the mid-18th century to modern times. Dwells on Oersted's work in both physics and chemistry, and his considerable influence in this area. L. Rosenfeld (*Centaurus* 11 (1965): 316-317) does not think the stated aim of investigating the social aspects of this development has been fulfilled.

276. [Leipzig University]. *Bedeutende Gelehrte in Leipzig: Zur 800-Jahr-Feier der Stadt Leipzig im Auftrag von Rektor und Senat der Karl Marx Universität heraus-*

gegeben. 2 vols. Vol. 1 ed. Max Steinmetz; vol. 2
ed. Gerhard Harig. Leipzig: Karl Marx Universität,
1965. xxv + 303 pp; ix + 211 pp.

A collection of short biographies of scholars and
scientists connected with the university, Vol. 1 being
devoted to "humanists," Vol. 2 to scientists. The
quality of the contributions is uneven, some being
well researched and with good bibliographies, while a
few are overly sparse in both content and documentation.

277.   Lindroth, Sten, *Kungl. Svenska Vetenskapsakademiens*
        Historia 1739-1818. 2 vols. in 3. Stockholm:
        Kungl. Vetenskapsakademien, 1967.

        A monumental piece of scholarship that includes ac-
        counts of the founding and administration of the Academy,
        its relations with other scientific academies, with
        Swedish scientific instrument makers and with wider
        Swedish society, its publications, and the scientific
        work of its members during the period stated. J.C.
        Wilcke's work in experimental physics is described at
        length, as is, somewhat more briefly, that of his suc-
        cessors.

277a.   Lyons, Henry, *The Royal Society, 1660-1940: A History*
        of its Administration under its Charters. Cambridge:
        at the University Press, 1944. x + 354 pp.

        A very thorough account within the limited objective
        of "record[ing] how the Society's Councils have adminis-
        tered its affairs" during the period stated.

278.   Maindron, Ernest, *L'Académie des Sciences: histoire de*
        L'Académie; fondation de l'Institut National;
        Bonaparte membre de l'Institut National. Paris: F.
        Alcan, 1888. 344 pp.

        The best of the three "standard" 19th-century histories
        of the Academy as it provides a thorough account of
        references.

279.   Maury, Louis-Ferdinand-Alfred, *Les académies d'autrefois:*
        l'ancienne Académie des Sciences. Paris: Didier,
        1864.

        One of the "standard" histories of the Academy, es-
        pecially good for the 18th century, concentrating on the
        achievements of the society through its publications.
        Unfortunately undocumented.

280. Ostrovityanov, K.V., ed., *Istoriya Akademii Nauk SSSR*.
      Moscow/Leningrad: Izdatel'stvo Akademii Nauk SSSR,
      1958- .

      A straightforward institutional history, including
      summaries of the work done in the Academy in the various
      sciences in different periods. Vol. 1 covers the period
      1724-1803, Vol. 2 the period 1803-1917. A third volume
      is planned.

281. Schimank, Hans, "Zur Geschichte der Physik an der
      Universität Göttingen vor Wilhelm Weber (1734-1839)."
      Rete 2 (1974): 207-252.

      Chiefly concerned with detailing the career of Samuel
      Christian Hollmann (1694-1787), the first teacher of
      physics at the University of Göttingen. Also includes
      valuable comments on the meanings of the terms "mathe-
      matics" and "physics" at this period and general informa-
      tion about the University of Göttingen during the first
      fifty years of its existence.

282. Schmidt-Schönbeck, Charlotte, *300 Jahre Physik und
      Astronomie an der Kieler Universität*. Kiel: Ferdinand
      Hirt, 1965. 261 pp.

      The book is divided into four sections describing
      (i) the four main phases of the university's develop-
      ment; (ii) astronomy and physics in the 17th and 18th
      centuries; and (iii) and (iv), astronomy and physics up
      to the present day. At all times the teaching and re-
      search at the University is related to that in the wider
      scientific community, and although there is not too much
      scientific detail, there is considerable biographical
      information concerning individual physicists at Kiel
      including Hertz, Planck, Lenard and Pfaff. A good bib-
      liography.

283. Schofield, Robert E., "Histories of Scientific Societies:
      Needs and Opportunities for Research," *Hist. Sci.* 2
      (1963): 70-83.

      A critical survey of material available on scientific
      societies--especially in Britain--from the 17th to the
      19th century, highlighting areas that need further study.

284. Sedillot, Louis Pierre Eugene Amélie, "Les professeurs
      de mathématiques et de physique générale au Collège
      de France." *Bollettino di bibliografia e di storia
      delle scienze matematiche e fisiche* 2 (1869): 343-368,

387–448, 461–510; 3 (1870): 107–170. Reprinted, Rome:
Imprimerie des sciences mathématiques et physiques,
1869. 204 pp.

A well documented but fairly sketchy administrative
history of the College from 1530 to 1869 including accounts of the professional lives and works of the professors.

285. Shinn, Terry, *Savoir scientifique et pouvoir social:
      l'Ecole Polytechnique, 1794-1914*. Paris: Presses de
      la fondation nationale des sciences politiques, 1980.
      260 pp.

A study of the place of the Ecole Polytechnique within
French society during the period stated, incorporating
much statistical detail on the social backgrounds and
careers of the *polytechniciens*.

285a. Weld, Charles Richard, *A History of the Royal Society,
      with Memoirs of the Presidents, compiled from Authentic
      Documents*. 2 vols. London: John W. Parker, 1848.
      xx + 527 pp.; viii + 611 pp.

Still the only attempt at a comprehensive history of
the Society, but deals only with the period up to 1830.

(c) Instrumentation

286. Bonelli, Maria Luisa, *Il Museo di Storia della Scienza
      a Firenze*. Milan: Electra, 1968. 252 pp.

Contains 143 plates illustrating over 200 items and
a catalogue of the instruments in the collection, unfortunately not cross-referenced to the illustrations.
Includes many items of the 18th and 19th centuries.
Reviewed by G.L'E. Turner, *Brit. J. Hist. Sci.* 6
(1972-73): 206.

287. Bradbury, Savile, *The Evolution of the Microscope*.
      Oxford: Pergamon Press, 1967. x + 357 pp.

A well illustrated work which recounts the history
of the microscope from earliest times to the modern
day, concentrating on the period 1650-1900. Covers the
development of simple, compound, and achromatic instruments, and the technical advances of the Victorian era.

Reviewed by T.R. Levere, *Ann. Sci.* 24 (1968): 342-344.
An abridgement has also been published entitled *The
Microscope, Past and Present* (1968).

288. Bradbury, Savile, and G.L'E. Turner, eds., *Historical
Aspects of Microscopy.* Cambridge: Heffer, 1967. 227
pp. Also in *Proc. Roy. Microscopical Soc.* 2(1) (1967).

Contains, *inter alia*, an article by Bradbury assessing
the image quality achieved by a variety of microscopes
from 1700 to 1840, and another by Turner similarly
assessing the resolving power of microscopes from *c.*
1830 to 1900.

289. Brown, Joyce, "Guild Organization and the Instrument-
Making Trade, 1550-1830: The Grocers' and Clockmakers'
Companies." *Ann. Sci.* 36 (1979): 1-34.

Traces briefly the history of the two guilds with regard to their incorporation of instrument makers, showing
the decline in their membership and influence during
the 18th and 19th centuries. Gives a tentative list of
instrument makers in the Clockmakers Company, 1631-1730.

290. Brown, Joyce, *Mathematical Instrument-Makers in the
Grocers' Company 1688-1800.* London: The Science
Museum, 1979. xv + 103 pp.

A valuable history of instrument makers registered
with the Grocers' Company, including sections on statistics of the trade in the Company, a listing of "master
makers" and a further listing of "distinguished makers."
Good bibliography and index.

291. Bryden, D.J., *Scottish Scientific Instrument Makers
1600-1900.* Edinburgh: H.M.S.O. for the Royal Scottish
Museum, 1972. x + 59 pp.

An excellent account combining a social analysis of
the trade and an account of its development with a list
of Scottish makers and their dates, addresses and manu-
facturing specialities.

292. Davies, A.C., "The Life and Death of a Scientific Instrument: The Marine Chronometer, 1770-1920." *Ann. Sci.*
35 (1978): 509-525.

Traces the rapid expansion and then the slow decline
of the British chronometer industry, arguing that the
decline was due to the use of craft rather than factory
methods of production, and later to oversupply.

293. Disney, Alfred N., C.F. Hill and W.E.W. Baker, eds.,
      Origin and Development of the Microscope, as Illus-
      trated by Catalogues of the Instruments and Accessories
      in the Collection of the Royal Microscopical Society.
      London: Royal Microscopical Society, 1928. xii +
      303 pp.

      The latter half of the book is concerned with brief
      descriptions of about 160 instruments, in roughly chrono-
      logical order, from the 1670s to the end of the 19th cen-
      tury. Most of the descriptions of pre-1850 devices are
      accompanied by a photograph.

294. Goodison, N., *English Barometers 1680-1860: A History*
      of Domestic Barometers and their Makers. London:
      Cassell, 1969. xiv + 353 pp.

      A well illustrated work divided into three parts: the
      first describes the instruments and their chronological
      development; the second provides biographical details
      of "about fifty important makers" illustrated by samples
      of their work; and the third lists about 1700 other
      makers with brief, coded descriptions of many of their
      instruments.

295. Lavèn, W.J., and J.G. van Cittert-Eymers, *Electrostatical*
      Instruments in the Utrecht University Museum. Utrecht:
      Utrecht University Museum, 1967. 78 pp.

      Provides individual descriptions of over 120 instruments
      in the collection dating from the early 18th to the late
      19th century. These are grouped according to type of
      instrument, with each section having a short introduc-
      tion. Includes 49 photographs.

296. Michel, H., "La baromètre liégeois." *Physis* 3 (1961):
      205-212.

      Describes and analyzes the mathematics of a water
      barometer popular from the 17th to the 19th century.

297. Middleton, W.E. Knowles, *The History of the Barometer.*
      Baltimore: Johns Hopkins University Press, 1964. xx +
      489 pp.

      Meticulously researched and highly readable history
      which traces the chronological development of the
      barometer as an instrument. Subdivided into sections
      on the several different types, which are placed in the
      scientific context of the time. Includes much valuable

information on instrument makers, plentiful quotations
(translated) from primary sources, and approximately
200 accurate and informative figures.

298. Middleton, W.E. Knowles, *A History of the Thermometer
and Its Use in Meteorology*. Baltimore: Johns Hopkins
University Press/London: Oxford University Press,
1966. xiii + 249 pp.

A standard text, tracing the history of the different
kinds of thermometers and their uses in science. As
the author remarks, the history is "largely an account
of attempts to make and justify choices" of scales and
thermometric substances. Based on close and frequent
references to sources, presented here in translation,
and well illustrated with both line drawings and photo-
graphs. Reviewed by H.G. Korber, *Isis* 59 (1968): 214-
216 and D. Chilton, *Brit. J. Hist. Sci.* 4 (1968-69):
179.

299. Pipping, Gunnar, *The Chamber of Physics: Instruments in
the History of Sciences Collections of the Royal
Swedish Academy of Sciences, Stockholm*. Stockholm:
Almqvist & Wiksell, 1977. 250 pp.

A detailed catalogue of a major collection of astronomical, surveying and physical apparatus. Particular
attention is paid to the equipment assembled by J.C.
Wilcke during his long career (1759-1796) as the Academy's
Thamian Lecturer in Experimental Physics, and to the
development of scientific instrument making in Sweden.

300. Turner, G.L'E., *Essays on the History of the Microscope*.
Oxford: Senecio Publishing Co., 1980. 245 pp. 66
figs.

A very useful collection of 12 of the author's previously published papers on aspects of the history of
the microscope.

301. Turner, G.L'E., "A History of Optical Instruments: A
Brief Survey of Sources and Modern Studies." *Hist.
Sci.* 8 (1969): 53-93. Reprinted, *Essays on the History
of the Microscope* by G.L'E. Turner. Oxford: Senecio
Publishing Co., 1980, pp. 31-72.

A pithy bibliographical essay listing the most important
works on collections and inventories, histories of optical glass, the microscope and the telescope, and their
makers. Concludes with an extensive bibliography.

302.   Turner, G.L'E., "Micrographia historica: The Study of
       the History of the Microscope." *Proc. Roy. Micro-
       scopical Soc.* 7 (1972): 120-149. Reprinted, *Essays
       on the History of the Microscope* by G.L'E. Turner.
       Oxford: Senecio Publishing Co., 1980, pp. 1-29.

       Surveys previous historical writings on microscopy,
       which are seen to have been written chiefly from an
       antiquarian point of view. Argues that the historical
       study of scientific instruments such as the microscope
       can lead to wider and more significant historical con-
       clusions.

       (d) Foundational Issues, Properties of Matter, Sound

303.   Agassi, Joseph, "Leibniz's Place in the History of
       Physics." *J. Hist. Ideas* 30 (1969): 331-344.

       Concerned with "the roots of Einstein's Leibnizianism"
       with respect to the relational view of space. Takes
       it as read that "Einstein was not directly influenced
       by the works of Leibniz" and seeks to trace that in-
       fluence instead through intervening 18th and 19th-
       century writers concerned with the nature of space,
       especially Kant and Boscovich.

304.   Harvey, E. Newton, *A History of Luminescence from the
       Earliest Times until 1900*. Philadelphia: American
       Philosophical Society, 1957. xxiv + 692 pp.

       An encyclopaedic study that includes extensive dis-
       cussions of many 18th and 19th-century investigations
       of luminous phenomena of all kinds.

305.   Jammer, Max, *Concepts of Mass in Classical and Modern
       Physics*. Cambridge, Mass.: Harvard University Press,
       1961. 230 pp.

       An excellent work. Among 18th-century authors, Kant
       and Euler receive most attention. Nineteenth-century
       contributions are discussed in chapters on the "modern"
       (i.e., Mach-ian), the gravitational and the electro-
       magnetic concepts of mass.

306.   Jammer, Max, *Concepts of Space: The History of Theories
       of Space in Physics*. Cambridge, Mass.: Harvard Uni-
       versity Press, 1954. xvi + 196 pp.

Describes Newton's concept of absolute space, with criticisms of it by such writers as Kant, Lange and Mach.

307. Lindsay, Robert B., "Historical Introduction." *The Theory of Sound*, by John William Strutt, third Baron Rayleigh. New York: Dover, 1945. Vol. 1, pp. v-xxxii.

One of the few worthwhile studies on the history of sound. Includes a biographical sketch of Rayleigh, a survey of the "historical development of acoustics to the time of Rayleigh," and a summary of Rayleigh's contributions to the subject.

308. Miller, Dayton C., *Anecdotal History of the Science of Sound to the Beginning of the 20th Century*. New York: Macmillan, 1935. xiv + 114 pp.

One of the few works on this subject, by a leading 20th-century investigator. Consists of brief paragraphs describing the work of one investigator after another, without any pretense of historical synthesis.

309. Scott, Wilson L., *The Conflict between Atomism and Conservation Theory, 1644 to 1860*. London: Macdonald/ New York: Elsevier, 1970. xiv + 312 pp.

Book I describes the attempts to reconcile theories of mechanics with the idea of hard inelastic atoms, the latter being seen as philosophically necessary but requiring, impossibly, infinite forces in inelastic collisions. Book II describes the confused evolution of the notions of transformation of *vis viva* to work in such collisions and conservation of *vis viva* and energy, showing the gradual replacement of the idea of inelastic atoms with that of elastic atoms. Deals chiefly with the work of Newton, Maclaurin, d'Alembert, Maupertuis, the Carnots, Lagrange and Poisson.

310. Scott, Wilson L., "The Significance of 'Hard Bodies' in the History of Scientific Thought." *Isis* 50 (1959): 199-210.

A summary of some of the ideas set out in his book published some years later (item 309).

311. Timoshenko, Stephen, *History of the Strength of Materials, with a Brief Account of the History of the Theory of Elasticity and Theory of Structures*. New York, Toronto and London: McGraw-Hill, 1953. x + 452 pp.

A broad treatment of the field which of necessity
leaves out considerable detail and is somewhat selective of material, especially in pre-1800 work; two
thirds of the book is devoted to the 19th century.
Breaks up the subject into sections on the mathematical
theory of elasticity and more empirical engineering
work on the strength of materials, chronologically arranged with a strong emphasis on the biography of individuals. Uses modern mathematical notation with plenty
of diagrams.

312.   Todhunter, Isaac, *History of the Theory of Elasticity*
       and of the Strength of Materials from Galilei to Lord
       Kelvin. Edited and completed by Karl Pearson. 2 vols.
       Cambridge: 1886-93. Reprinted, New York: Dover/London:
       Constable, 1960.

A valuable reference work which suffers somewhat from
a conflict of styles, having been completed by Pearson
(who wrote most of vol. 2) after Todhunter's death.
The over-brief section dealing with the period prior to
1800 is poor, but the period from 1800 to the 1860s is
treated quite comprehensively. The work is essentially
a string of scientific biographies, sometimes grouped
under subject headings. The bulk of the work is descriptive analysis of published memoirs, with plentiful cross
references but little historical analysis. Large sections on Poisson, Cauchy, Lamé, Saint-Venant, Kirchhoff,
Clebsch, Boussinesq and Kelvin amongst others. Has a
good index.

(e) Mechanics, Fluid Mechanics

313.   Aronson, Samuel, "The Gravitational Theory of Georges-
       Louis Le Sage." *Natural Philosopher* 3 (1964): 53-74.

Gives a clear account of Le Sage's ideas, their consolidation into a coherent mathematical theory by Preston
and Kelvin almost 100 years afterwards, and the demo-
lition of this theory by Maxwell on thermodynamic grounds.

314.   Biswas, Asit K., *History of Hydrology*. London: North
       Holland/New York: American Elsevier, 1970. xii + 336 pp.

Necessarily very selective, since it examines material
from c. 600 B.C. to 1900 A.D. Provides a roughly chronological account of various individuals and their discoveries

315. Dircks, Henry, *Perpetuum mobile, or Search for Self-Motive Power During the 17th, 18th and 19th Centuries: Three Centuries of Perpetual Motion*. London: 1861. xlii + 558 pp.

With extensive quotations from a large variety of different (though mainly British) sources, describes numerous supposed perpetual motion devices and a host of publications, including many by prominent physicists, some in favour of and some opposing the theory of perpetual motion.

316. Dugas, René, *A History of Mechanics*. Translated by J.R. Maddox. Neuchâtel: Editions du Griffon/New York: Central Book Company, 1955. 671 pp.

A translation of the author's *Histoire de la mécanique* (1950). Traces the evolution of mechanics from Aristotle to the 1930s. Parts 3 and 4 respectively deal with 18th and 19th-century classical physics. Provides brief analyses of major works of individual mechanicians, not really placed in the context of other contemporary work or even of work by the same individual. Reviewed by I.B. Cohen, *Isis* 42 (1951): 271-272.

317. Flachsbart, O., "Geschichte der Experimentellen Hydro- und Aeromechanik, insbesondere der Widerstandsforschung." *Handbuch der Experimentalphysik*, ed. W. Wien and F. Harms. Leipzig, 1926-35. Vol. 4(2), pp. 3-61.

An old-style survey of contributions to the science in question, chiefly concentrating on 18th and 19th-century work.

318. Grigor'ian, A.T., *Mekhanika v Rossii*. Moscow: Izdatel'stvo "Nauka," 1978. 192 pp.

A broad survey of Russian contributions to the science of mechanics from the founding of the St. Petersburg Academy of Sciences to the 1917 Revolution.

319. Mach, Ernst, *The Science of Mechanics: A Critical and Historical Account of Its Development*. Translated by Thomas J. McCormack. 6th ed., with revisions through the 9th German ed. La Salle, Ill.: Open Court Publishing Co., 1960. xxxi + 634 pp.

First published, in German, in 1883. A work most famous for its critique of the concept of mass and of Newton's ideas of absolute space and absolute time. Chiefly concerned with the foundations of the science,

and thus, so far as historical exposition goes, with
developments up to and including the work of Newton.
Does, however, include some discussion of later (mostly
18th-century) work in chapters on "The Extended Application of the Principles of Mechanics and the Deductive
Development of the Science" and "The Formal Development
of Mechanics." The final chapter includes some provo-
cative remarks on "The Relations of Mechanics to Physics."

320. Neményi, P.F., "The Main Concepts and Ideas of Fluid
      Dynamics in their Historical Development." *Arch. Hist.
      Exact Sci.* 2 (1962-66): 52-86.

      Briefly traces the development of concepts such as
      fluid flow, resistance and turbulence from Aristotle to
      the end of the 19th century.

321. Rouse, Hunter, and Simon Ince, *History of Hydraulics.*
      Iowa City: State University of Iowa, 1957. Reprinted
      New York: Dover/London: Constable, 1963. xii + 269 pp.

      Has as its aim the description of the formulation of
      the underlying principles of fluid motion. The treatment
      of each aspect is necessarily brief. C.A. Truesdell
      (review, *Isis* 50 (1959): 69-71) criticizes the work for
      its obvious reliance on flawed secondary sources and
      its undervaluation of 18th-century theoretical work,
      especially of the Basle school. While descriptions of
      the mathematical theory are generally sketchy, the
      engineering and empirical aspects of the development of
      practical hydraulics are clearly and concisely explained.

322. Saint-Venant, A.J.C. Barré de, and C.L.H. Navier, *Resumé
      des leçons données à l'Ecole des Ponts et Chaussées
      sur l'application de la mécanique à l'établissement
      des constructions et des machines.* 3rd ed. Vol. 1.
      With notes and appendices by A.J.C. Barré de Saint-
      Venant. Paris: Dunod, 1864. cccxi + 852 pp.

      The long introduction and appendices provide a valuable
      source for the history of elasticity theory 1750-1850,
      but must be used with caution. Subjects Green, Stokes
      and Maxwell to severe and unwarranted criticism.

*   Scott, Wilson L., *The Conflict between Atomism and Con-
      servation Theory, 1644 to 1860.*

      Cited herein as item 309.

323. Szabó, István, *Geschichte der mechanischen Prinzipien
      und ihrer wichtigsten Anwendungen*. Basle/Stuttgart:
      Birkhäuser, 1976. xvi + 491 pp. (Wissenschaft und
      Kultur, 32).

      Aiming to cover the period 1600–1975, the author has
      divided the book into five chapters: the earliest es-
      tablishment of the classical mechanics of rigid bodies
      (Newton, Euler, d'Alembert); the development of mechanical
      principles from the 17th to the 19th century; fluid
      mechanics; the linear theory of homogeneous and isotropic
      elastic materials; and impact theory. Of necessity the
      topics are selective but close attention is paid to
      original writings. Reviewed by E.A. Fellman, *Isis* 70
      (1979): 469–471 and C.A. Truesdell, *Centaurus* 23 (1980):
      163–175.

*   Timoshenko, Stephen, *History of the Strength of Materials,
      with a Brief Account of the History of the Theory of
      Elasticity and Theory of Structures*.

      Cited herein as item 311.

324. Todhunter, Isaac, *A History of Mathematical Theories of
      Attraction and the Figure of the Earth from the Time
      of Newton to that of Laplace*. 2 vols. London: Mac-
      millan, 1873. Reprinted, 2 vols. in 1, New York: Dover/
      London: Constable, 1962.

      The classic study of the subject. Provides critical
      technical discussions of the original papers and treatises
      on the subject, with occasional brief discussions on
      the evolution of mathematical techniques, all in roughly
      chronological order. Most attention is focussed on
      Legendre and especially Laplace with an attempt to es-
      tablish the latter's precursors. The mathematical sym-
      bols have been regularized and the bounds of the work,
      in fact, extend to 1825.

*   Todhunter, Isaac, *History of the Theory of Elasticity
      and of the Strength of Materials from Galilei to Lord
      Kelvin*.

      Cited herein as item 312.

325. Tokaty, G.A., *A History and Philosophy of Fluid Mechanics*.
      Henley-on-Thames, England: G.T. Foulis, 1971. x +
      241 pp.

      Follows the format of a series of short discussions of
      individual scientists and their work. Initially these

are organized into a rough history of the development
of the mathematical equations of fluid flow (Newton to
Stokes and Navier in twenty-six pages) but this section
is followed by what seems to be a random series of observations on the mathematical, experimental and (especially) engineering aspects of fluid mechanics in its
broadest sense, from the 18th to the 20th century. The
"philosophy" of the title is restricted to a few maxims
on theory and experiment inserted at random.

326.   Truesdell, Clifford A., "Notes on the History of the
       General Equations of Hydrodynamics." *Amer. Math.
       Monthly* 60 (1953): 445-458.

An introductory general history which briefly describes
the major contributions to the subject by Newton, D. Bernoulli, d'Alembert, Euler, Navier, Cauchy, Poisson and
Stokes.

(f) Light

327.   Halbertsma, K.T.A., *A History of the Theory of Colour*.
       Amsterdam: Swets and Zeitlinger, 1949. 267 pp.

Describes several fragmented efforts to evolve a theory
of colour, involving aspects of physics, mathematics and
physiology. The section on the 18th century is largely
concerned with Goethe's violently anti-Newtonian work.
19th-century work discussed includes that of Young,
Voigt and especially Helmholtz, in the areas of sensa-
tion, mixing, and intensity. Shows that a desire to
evolve a mathematical theory of colour was not enough:
it was also necessary to develop an adequate theory of
perception.

328.   Hanson, N.R., "Waves, Particles, and Newton's 'Fits.'"
       J. Hist. Ideas 21 (1960): 370-391.

Briefly describes the evolution of the view that the
wave and particle theories of light were incompatible.
Argues that Newton's theory of fits could have accommo-
dated the findings of Young and Fresnel, and that with
hindsight we can see that it was Foucault's experiment
on the velocity of light in different media that was the
"crucial" refutation of Newton.

* Harvey, E. Newton, *A History of Luminescence from the Earliest Times until 1900.*

  Cited herein as item 304.

329. Hoppe, Edmund, *Geschichte der Optik.* Leipzig: J.J. Weber, 1926. Reprinted, Wiesbaden: M. Sandig, 1967. ix + 263 pp.

  A general survey, the greater part of which deals with the period from Newton to the end of the 19th century.

* Laudan, Larry, "The Epistemology of Light: Some Methodological Issues in the Subtle Fluids Debate."

  Cited herein in item 255.

330. Mach, Ernst, *The Principles of Physical Optics: An Historical and Philosophical Treatment.* London: Methuen, 1926. xii + 324 pp.

  First published in German, posthumously, in 1921. Written in a similar style to Mach's earlier studies of mechanics and heat (items 319, 344) with the intention of laying bare "the origin of the general concepts of optics and the historical threads in their development, extricated from metaphysical ballast."

331. Ronchi, Vasco, *The Nature of Light: An Historical Survey.* London: Heinemann, 1970. xii + 288 pp.

  First published in Italian as *Storia della luce* (1939). Chaps. 6 and 7 (pp. 209-259) are devoted to the 18th and 19th centuries, and include much valuable information. The approach adopted is, however, very positivistic.

* Whittaker, Edmund Taylor, *A History of the Theories of Aether and Electricity.*

  Cited herein as item 349a. Includes extensive discussions of 18th and 19th-century views of the "luminiferous medium."

332. Worrall, John, "The Pressure of Light: The Strange Case of the Vacillating 'Crucial Experiment.'" *Studs. Hist. Phil. Sci.* 13 (1982): 133-171.

  Surveys various attempts since the early 18th century to detect radiation pressure, and also the different attitudes adopted at different times as to the significance

of the results obtained--sometimes positive, sometimes
negative--for the opposing "wave" and "corpuscle"
theories of light.

(g) Heat

*   Cardwell, D.S.L., *Technology, Science and History*.

         Cited herein as item 239.

333.   Fox, Robert, *The Caloric Theory of Gases from Lavoisier
         to Regnault*. Oxford: Clarendon Press/New York: Oxford
         University Press, 1971. xvi + 371 pp.

         Incorporates extensive discussion on caloric theories
         of heat and related theories of the nature of light.
         After briefly examining 17th and 18th-century work, Fox
         concentrates on post-1780 caloric theories, especially
         that of Laplace and Lavoisier. He examines the role of
         caloric in Carnot's theory of the heat engine, in La-
         place's determination of the speed of sound, and in
         explaining the relationship between heat and light.
         Finally he describes the decline of caloric theory,
         attributing this to a more general rejection of La-
         placian physics and the emergence of positivist and
         skeptical attitudes. A good primary and secondary bib-
         liography. Reviewed by S. Pierson, *Isis* 68 (1977):
         462-464.

334.   Mach, Ernst, *Die Principien der Wärmelehre: Historisch-
         kritisch entwickelt*. Leipzig, 1896. viii + 472 pp.
         4th ed., 1923.

         A classic study of the historical development of the
         science of heat, written from Mach's well-known critical
         philosophical perspective.

335.   Meyer, Kirstine, *Die Entwicklung des Temperaturbegriffes
         im Laufe der Zeiten*. Braunschweig, 1913. Reprinted,
         New York: Arno Press, 1981. vi + 160 pp.

         The definitive study of the concept of temperature
         and its measurement.

336.   Roller, Duane, *The Early Development of the Concepts of
         Temperature and Heat: The Rise and Decline of the
         Caloric Theory*. Cambridge, Mass.: Harvard University

Press, 1950. iv + 106 pp. (Harvard Case Histories in Experimental Science, 3).

A good though now dated introduction to the field, incorporating extensive quotations from original sources. Describes the experimental and theoretical work of Black, Rumford and Davy.

337. Truesdell, C.A., "Early Kinetic Theories of Gases." *Arch. Hist. Exact. Sci.* 15 (1975-76): 1-66.

A study of kinetic theories from the eighteenth century up to Maxwell's first theory of 1860. The emphasis is on summarizing and evaluating results obtained. Includes an extensive critical and chronological bibliography of primary sources.

(h) Electricity and Magnetism

338. Bauer, Edmond, *L'Electromagnétisme hier et aujourd'hui.* Paris: Albin Michel, 1949. 348 pp.

Includes a now rather dated section on electrical theory and experiment from the eighteenth to the early twentieth century (pp. 39-139), and a useful discussion of Coulomb's work (pp. 213-235).

339. Carr, Margaret E.J., "The Development of Mathematical Theories of Electricity prior to the Work of Maxwell, with Special Reference to the Concept of Potential." M.Sc. thesis, University of London (University College), 1949. 203 pp.

Gives useful summaries of the work of leading contributors to the field.

340. Ekelöf, Stig., ed., *Catalogue of Books and Papers relating to the History of Electricity in the Library of the Institute for Theoretical Electricity, Chalmers University of Technology.* 2 vols. Göteborg, 1964-66. 109 pp.; 111 pp.

The catalogue of the valuable collection that has been formed over many years by Professor Ekelöf. Vol. I deals with works published prior to 1820, Vol. II with those published after that date. The collection is rich in Continental and especially Scandinavian material. It

does not aspire to completeness; the aim has been to
assemble "not all works, but all *important* works."

341. Finn, Bernard S., "History of Electrical Technology: The
      State of the Art." *Isis* 67 (1976): 31-35.

      Bibliographical essay, mainly useful in the present
      context for the writings on 18th-century electrical ap-
      paratus, pp. 31-32.

342. Frost, A.J., ed., *Catalogue of Books and Papers relating
      to Electricity, Magnetism, the Electric Telegraph, etc.,
      including the Ronalds Library*. London, 1880. xxvii +
      564 pp.

      A catalogue initially compiled over many years by the
      famous telegraph engineer, Sir Francis Ronalds, F.R.S.,
      up to his death in 1873. The catalogue includes not
      only titles of the works in Ronalds' own remarkable
      collection of materials on electricity, magnetism and
      related subjects, but also all other works on the same
      subjects which came to his notice. Entries are listed
      alphabetically by author. The Ronalds collection it-
      self, particularly rich in Continental publications, is
      now housed at the Institute of Electrical Engineers,
      London, where it is supplemented by the smaller but also
      very valuable Silvanus P. Thompson collection.

343. Gartrell, Ellen G., *Electricity, Magnetism and Animal
      Electricity: A Checklist of Printed Sources, 1600-
      1850*. Wilmington, Del.: Scholarly Resources, 1975.
      x + 125 pp.

      Works are catalogued alphabetically with separate
      sections for Electricity and Magnetism (pp. 1-85, 1029
      items), and Animal Magnetism (pp. 87-105, 213 items).
      Includes subject and author index.

344. Hoppe, Edmund, *Geschichte der Elektrizität*. Leipzig:
      J.A. Barth, 1884. xx + 622 pp.

      A thorough and very comprehensive general history of
      electricity up to the discovery of conservation of
      energy and Weber's electric force "law." Concludes
      with a substantial section on 19th-century technical
      applications.

345. Lang, Sidney B., *Source Book of Pyroelectricity*. London,
      New York, Paris: Gordon and Breach Science Publishers,
      1974. xv + 562 pp.

The best guide to the subject, including a very useful
survey of the literature on pyroelectricity from the
Greeks to the end of the nineteenth century, along with
a large number of illustrations of various types of
apparatus.

346.   Marton, L., and C. Marton, "Evolution of the Concept of
        Elementary Charge." *Advances in Electronics and Elec-
        tron Physics* 50 (1980): 449-472.

        Traces the history of the concept of elementary charge,
        arguing for the elimination of Franklin and Aepinus as
        originators of the concept. Describes the speculations
        of (among others) Faraday, Maxwell, Stoney, Helmholtz
        and Zeeman, and stresses the importance of Lorentz' work.
        Finishes with a discussion of the concepts of "discovery"
        and "discoverer," emphasizing their historical nature
        and seemingly random application.

347.   Mottelay, Paul Fleury, *Bibliographical History of Elec-
        tricity and Magnetism, Chronologically Arranged*....
        London: Charles Griffin & Co., 1922. xx + 673 pp.

        A monumental work, chronologically arranged, and ad-
        vancing from earliest times up to the age of Faraday.
        Entries amount to brief accounts of the publications,
        discoveries or inventions attributed to the authors noted
        at each date, to which are appended more or less extensive
        lists of authorities.

348.   Teichmann, Jürgen, *Zur entwicklung von Grundbegriffen
        der Elektrizitätslehre, insbesondere des elektrischen
        Stromes bis 1820*. Hildesheim: H.A. Gerstenberg,
        1974. viii + 162 pp.

        Describes some of the difficulties experienced in
        developing the fundamental concepts of current electricity
        --voltage, capacity, resistance, current--during the
        period 1771-1820. Concentrates on the work of Beccaria,
        Volta, Cavendish and Ritter.

349.   Weaver, William D., ed., *Catalogue of the Wheeler Gift
        of Books, Pamphlets and Periodicals in the Library of
        the American Institute of Electrical Engineers*. With
        introduction, descriptive and critical notes by Brother
        Potamian. 2 vols. New York, 1909. viii + 504 pp.;
        475 pp.

        An annotated catalogue of the celebrated Latimer Clark
        collection, of which it was claimed at the time it was

sold that it held "practically every known publication
in the English language previous to 1886, on magnetism,
electricity, galvanism, the lodestone, mariner's compass,
etc." A number of other items are included that were
published after that date, and many in other languages.
The order of entries in the catalogue is chronological,
with a separate chronological listing of "Excerpts from
Periodicals-Miscellanea." There is an index of authors
and an elaborate system of cross-referencing between
entries. There is also a separate subject index for
the very extensive section of the collection that relates
to the telegraph.

349a. Whittaker, Edmund Taylor, *A History of the Theories of
Aether and Electricity*. Vol. I: *The Classical Theories*.
Vol. II: *The Modern Theories 1900-1926*. London: Thomas
Nelson & Sons, 1951-53. Reprinted, New York: Harper
Torchbooks, 1960.

Vol. I is a revised edition of a work first published
in 1910 under the title *A History of the Theories of
Aether and Electricity, from the Age of Descartes to the
Close of the Nineteenth Century*. Still the standard
general history of the subject, and likely to remain so
even though recent historiography has challenged some of
its premises, and recent scholarship some of its conclusions. The treatment of Einstein in Vol. II is,
however, notoriously unfair.

V. EIGHTEENTH-CENTURY PHYSICS

(a) General

349b. Adickes, Erich, *Kant als Naturforscher*. 2 vols. Berlin:
W. de Gruyter & Co., 1924-25.

A systematic and very thorough analysis of Kant's
writings on various aspects of physics, ranging from the
notion of force and Kant's dynamical theory of matter,
through mechanics, the nature and functions of the
aether, the theory of the heavens, and the natural history of the Earth.

350. Barber, William H., "Mme du Châtelet and Leibnizianism:
The Genesis of the *Institutions de physique*." *The
Age of Enlightenment: Studies Presented to Theodore
Besterman*, ed. W.H. Barber et al. Edinburgh and Lon-
don: Oliver and Boyd, 1967, pp. 200-222.

Describes du Châtelet's process of self-education,
arguing that midway through the *Institutions* (1739) she
abruptly changed from Newtonian to Leibnizian metaphysics
while retaining Newton's physics.

351. Barber, William H., *Leibniz in France from Arnauld to
Voltaire: A Study in French Reactions to Leibnizianism,
1670-1760*. Oxford: Clarendon Press, 1955. xii +
276 pp.

While mainly concerned with wider philosophical issues--
notably that of Leibnizian optimism--Barber gives much
valuable information on the mostly less than enthusiastic
reception of Leibnizianism and Wolffianism in France.

351a. Beer, Peter, ed., *Newton and the Enlightenment: Proceedings of an International Symposium, held at Cagliari,
Italy, on 3-5 October 1977*. Oxford: Pergamon Press,
1978. (*Vistas in Astronomy*, vol. 22 (4)).

Includes the following (mostly quite short) papers:
A.R. Hall, "Newton--the Eighteenth Century's Marble
Image"; E.G. Forbes, "Newton's Science and the Newtonian
Philosophy"; J. Agassi, "The Ideological Import of New-
ton"; G. Solinas, "Newton and Buffon"; F. Restaino,
"Newton e la filosofia scozzese"; P. Casini, "R.G. Bosco-
vich and Newton's *Opticks*"; M.B. Hall, "Newton and his
Theory of Matter in the Eighteenth Century"; J.O. Fleck-
enstein, "Die Hermetische Tradition in der Kosmologie
Newtons"; G.A.J. Rogers, "Locke, Newton and the Enlighten-
ment"; K. Figala, "Pierre des Maizeaux's View of Newton's
Character"; M.A. Hoskin, "Newton and Lambert"; R. Taton,
"Sur la diffusion des théories newtoniennes en France:
Clairant et le problème de la terre"; A. Hayli, "Le grand
tournant de la pensée newtonienne de 1688-1690"; H.C.
Freiesleben, "Newton's Quadrant for Navigation"; I.B.
Cohen, "Notes on Newton in the Art and Architecture of
the Enlightenment"; S.J. Dundon, "The Trajectory of Es-
sentialism from Newton to d'Alembert"; M.C. Jacob, "New-
tonian Science and the Radical Enlightenment."

352. Berger, Peter, "Johann Heinrich Lamberts Bedeutung in der
      Naturwissenschaft des 18. Jahrhunderts." *Centaurus* 6
      (1959): 157-254.

      A detailed study of Lambert's philosophy of science.

353. Bernal, J.D., "Les rapports scientifiques entre la
      Grand-Bretagne et la France au XVIIIe siècle." *Rev.
      Hist. Sci.* 9 (1956): 289-300.

      Points out there was much less scientific contact be-
      tween France and Britain in the 18th than in the 17th
      century, attributing this to the general decrease in
      scientific impetus, to the differing political climates,
      and to the difference between amateur empirical British
      science and the theoretical institutionalized French
      variety. But also notes the value of such contact when
      it did occur as in the field of electricity and aspects
      of technology.

353a. Birembaut, Arthur, "Sur les lettres du physicien Magellan
      conservées aux Archives Nationales." *Rev. Hist. Sci.*
      9 (1956): 150-161.

      A preliminary account of 37 letters of J.H. de Magellan
      to Louis-Henri Duchesne spanning the period 1769-1774.
      The letters document Magellan's supplying the Parisian
      savants with the latest scientific news from London

together with books, instruments and even large pieces of machinery.

354. Boas, Marie, "Structure of Matter and Chemical Theory
      in the Seventeenth and Eighteenth Centuries." *Criti-
      cal Problems in the History of Science*, ed. Marshall
      Clagett. Madison: University of Wisconsin Press,
      1959, pp. 499-514.

      Examines "chemical" theories of matter during the
      period stated, showing that attraction became an integral
      part of particle theory during the 18th century, and that
      theorizing concentrated on discovering the smallest
      chemically active particles rather than the ultimate
      nature of matter.

355. Boss, Valentin, *Newton and Russia: The Early Influence,
      1698-1796*. Cambridge, Mass.: Harvard University Press,
      1972. xviii + 306 pp.

      A stimulating account of the reception of Newtonian
      science in Russia during the 18th century, and in par-
      ticular of Newtonian themes pursued (or opposed) within
      the St. Petersburg Academy of Sciences. Reviewed by
      A. Vucinich, *Isis* 65 (1974): 537-538.

356. Bowles, Geoffrey, "John Harris and the Powers of Matter."
      Ambix 22 (1975): 21-38.

      Argues for the influence of "non-scientific preoccu-
      pations" on the matter theory set out in Harris' *Lexicon
      Technicum*. In Vol. 1 (1704), Harris followed Boyle in
      presenting strictly corpuscularian ideas with direct
      interactions between corpuscles. In Vol. 2 (1710), he
      followed Newton and enthusiastically adopted a notion of
      "material forces" acting between particles.

357. Brockliss, L.W.B., "Aristotle, Descartes and the New
      Science: Natural Philosophy at the University of Paris,
      1600-1740." *Ann. Sci.* 38 (1981): 33-69.

      Details the decline of Aristotelian physics at the
      University of Paris on the basis of a systematic analysis
      of a large number of surviving lecture courses, student
      notebooks and theses from the 17th and early 18th
      centuries. Shows that Aristotelian ideas remained
      dominant until the 1690s, but were then replaced quite
      rapidly by a non-dogmatic probabilist form of Cartesian
      mechanical philosophy which continued to hold sway into
      the 1740s.

357a. Brockliss, L.W.B., "Philosophy Teaching in France, 1600-
        1740." *Hist. Univ.* 1 (1981): 131-168.

        Discusses, on the basis of a wide-ranging survey of
        surviving published and unpublished lecture courses, the
        evolution of the teaching of philosophy in the French
        collèges de plein exercise during the period stated.
        Argues that the traditional account, according to which
        the Aristotelian approach was replaced by that of Des-
        cartes in the second half of the seventeenth century,
        needs substantial modification. Only in the section of
        the course devoted to natural philosophy was this the
        case: here, Aristotelianism went on the defensive in
        the period 1640-1690, and was thereafter gradually sup-
        planted altogether. "In other respects, Aristotle
        weathered the Rationalist storm with relative ease."
        Suggests that the real impact of the Scientific Revolu-
        tion on the philosophy course lay in the gradual division
        of what had been an integrated subject into two isolated
        parts, the moral and natural sciences.

358.   Brunet, Pierre, *Les physiciens hollandais et la méthode
          expérimentale en France au XVIIIe siècle.* Paris:
          A. Blanchard, 1926. 153 pp.

        Contains two sections, the first dealing with the
        development of the experimental method in Holland and
        concentrating on the work and influence of 'sGravesande
        and Musschenbroek, the second concerned with the influ-
        ence of their work in France.

359.   Brunet, Pierre, *L'introduction des théories de Newton en
          France au XVIIIe siècle avant 1738.* Paris: A. Blan-
          chard, 1931. vii + 355 pp.

        A standard work describing the battle in France between
        Newton's theories and those of the Cartesians from 1700
        to 1738. A.R. Hall, "Newton in France: A New View,"
        History of Science 13 (1975): 233-50, is necessary
        additional reading, pointing out the limitations in the
        source material used by Brunet, his restriction of the
        debate primarily to celestial mechanics, and his ignoring
        of much contemporary work in France favourable to
        Newton.

360.   Brunet, Pierre, "L'oeuvre scientifique de Charles François
          Du Fay (1698-1739)." *Petrus nonius* 3(2) (1940): 1-19.

        A straightforward summary of Du Fay's principal publi-
        cations.

361. Brunet, Pierre, "Un grand débat sur la physique de Malebranche au XVIIIe siècle." *Isis* 20 (1934): 367-395.

Describes the debate among French natural philosophers
in the 1730s and 1740s between the defenders of Malebranche's "petits tourbillons," designed to replace the
spherical globules in Descartes' universe, and the
critics of the idea. The chief defender whose views are
discussed is Privat de Molières; the chief critics are
Banières and Sigorgne.

362. Buchdahl, Gerd, "Gravity and Intelligibility: Newton to
Kant." *The Methodological Heritage of Newton*, ed.
Robert E. Butts and John W. Davis. Oxford: Basil
Blackwell, 1970, pp. 74-102.

Includes extended discussions of the views of Newton
and Kant on the intelligibility or otherwise of action
at a distance--but, despite the title, virtually nothing
on developments in the intervening period.

362a. Calinger, Ronald S., "The Newtonian-Wolffian Confronta-
tion in the St. Petersburg Academy of Sciences (1725-
1746)." *Cah. Hist. Mond.* 11 (1968): 417-435.

Emphasizes the importance of Wolff's ideas and influence for 18th-century science, especially in the German
world and in the St. Petersburg Academy of Sciences.
Describes the debates over *vis viva* in the St. Petersburg
Academy during the period stated.

363. Calinger, Ronald S., "The Newtonian-Wolffian Controversy (1740-1759)." *J. Hist. Ideas* 30 (1969): 319-
330.

Describes a series of clashes at the Berlin Academy
between the generally Newtonian Maupertuis and Euler and
advocates of the Wolffian philosophy. The result was
a general though not complete acceptance of Newtonian
ideas.

364. Calinger, Ronald, "Euler's *Letters to a Princess of
Germany* as an Expression of his Mature Scientific
Outlook." *Arch. Hist. Exact Sci.* 15 (1975-76): 211-
233.

Presents Euler as an eclectic, espousing elements of
the Cartesian methodology and Newtonian mechanics and
empiricism, rejecting most Leibnizian ideas but also
Newtonian optics and matter theory, preferring a wave
theory of light and an ether.

364a. Calinger, Ronald, "Kant and Newtonian Science: The Pre-
      Critical Period." *Isis* 70 (1979): 349-362.

      Surveys Kant's writings on physics during his "pre-
      critical" period, arguing that "within a small learned
      circle that grew larger in the 1760s through correspon-
      dence and publications, Kant was a powerful native voice
      disseminating, criticizing, and selectively elaborating
      Newtonian science at the initial stage of its influence
      in Prussia."

364b. Carvalho, Joaquim de, "Correspondência científica dirigida
      a João Jacinto de Magalhães." *Revista da Faculdade de
      Ciências, Universidade de Coimbra* 20 (1951): 93-283.

      Publishes the texts of 54 letters, now preserved in
      the Bodleian Library, Oxford, addressed to Magellan,
      dating from the years 1769-1789, from a variety of cor-
      respondents in all parts of Europe.

365. Casini, Paolo, "Les débuts de newtonianisme en Italie,
      1700-1740." *Dix-huitième siècle* 10 (1978): 85-100.

      Argues that, although the evidence is sparse due to
      self-censorship by Italian intellectuals from fear of
      the Church, there was wide dissemination of Newtonian
      theories in Italy early in the 18th century. Populariza-
      tion was only achieved, however, by translation of works
      by Newtonians such as Derham (1728), Cheyne (1729) and
      Pemberton (1733), and especially through the work of
      Algarotti (1733).

366. Caverni, Raffaello, *Storia del metodo sperimentale in
      Italia*. 6 Vols. Florence: 1891-1900. Reprinted with
      an introductory note by G. Tabarroni, New York and
      London: Johnson Reprint, 1972. (Sources of Science,
      134).

      A valuable work left unfinished at the author's death.
      The treatment is by subject (including most aspects of
      physics and mechanics) and incorporates lengthy quota-
      tions from primary sources. Concentrates on the 17th
      century, but also incorporates much valuable material
      on 18th-century workers.

367. Christie, J.R.R., "The Origins and Development of the
      Scottish Scientific Community, 1680-1760." *Hist. Sci.*
      12 (1974): 122-141.

      Argues that the roots of Scotland's later scientific
      achievements lay in the earlier development of a scien-

tific community with common ideals based around the University of Edinburgh. This coincided with a decline in the political influence of the Church and the emergence of a group interested in "intellectual and economic improvement."

368. Christie, J.R.R., "The Rise and Fall of Scottish Science." *The Emergence of Science in Western Europe*, ed. M.P. Crosland. London: Macmillan, 1975, pp. 111-126.

Links the rise of Scottish science in the period of 1750-1800 to the post-Union political and economic ideals of improvement and independence of the ruling aristocratic elite, and its decline to the loss of power of this elite.

*   Cochrane, Rexmond C., "Francis Bacon and the Rise of the Mechanical Arts in 18th-Century England."

Cited herein as item 477.

369. Cohen, I. Bernard, "A Note concerning Diderot and Franklin." *Isis* 46 (1955): 268-272.

Despite never having met him, Diderot both (untypically) praised Franklin and summarized part of his work in his *L'interprétation de la nature* (1753). This is taken to show that he was sensitive to the most recent scientific discoveries.

370. Cohen, I. Bernard, *Franklin and Newton: An Inquiry into Speculative Newtonian Experimental Science and Franklin's Work in Electricity as an Example Thereof.* Philadelphia: American Philosophical Society, 1956. Reprinted, Cambridge, Mass.: Harvard University Press, 1966. xxvi + 657 pp.

A monumental pioneering study of Newton's influence on British (including American) and Dutch experimental physics during the 18th century, especially through his *Opticks* and the famous "Queries" appended to this work. Includes substantial discussions of the writings of Boerhaave, 'sGravesande, Desaguliers and Hales in particular, in addition to the history of 18th-century electricity that is the major preoccupation of the book. Here, Franklin's importance as an electrical theorizer and not merely the inventor of the lightning rod is securely established--indeed, according to more recent studies, perhaps somewhat overstated (cf. item 674).

370a. Costabel, Pierre, "La participation de Malebranche au
        mouvement scientifique--le modèle tourbillonaire."
        Malebranche: l'homme et l'oeuvre, 1638-1715. Paris:
        Vrin, 1967, pp. 75-101.

        Discusses (1) Malebranche's role in the development
        and dissemination of the infinitesimal calculus in France
        in the years after 1692, and (2) the gradual evolution
        during the period 1692-1712 of his views on the respec-
        tive roles of reason and experience in the establishing
        of physical truths, with experience gradually acquiring
        a larger place in his thinking.

371.   Dauben, J.W., "Marat: His Science and the French Revolu-
        tion." *Arch. Int. Hist. Sci.* 22 (1969): 235-261.

        After a discussion of Marat's contribution to elec-
        trical science, argues that the future revolutionary was
        not opposed to science, but, with some justification,
        to the institutional form in which it was encased in
        ancien régime France.

372.   Daumas, Maurice, "Precision of Measurement and Physical
        and Chemical Research in the Eighteenth Century."
        Scientific Change, ed. A.C. Crombie. London: Heine-
        mann, 1963, pp. 418-430.

        Argues that the desire for accurate quantitative data
        in physics and chemistry did not emerge until the end
        of the 18th century. In the same period makers first
        developed the technique essential to manufacture instru-
        ments of the required accuracy.

373.   Delorme, Suzanne, "La vie scientifique à l'époque de
        Fontenelle d'après les 'Eloges des Savants.'" *Archeion*
        19 (1937): 217-235.

        Briefly describes scientific life in Paris during the
        period 1699-1739, showing the influence on science of
        various social and political factors. Argues that while
        most physicists were Cartesian in outlook, there were
        some partisans of Newton.

373a. Dundon, S.J.S., "Philosophical Resistance to Newtonianism
        on the Continent, 1700-1760." Ph.D. thesis, St. John's
        University, 1972. 419 pp. (University Microfilms
        order no. 72-21719).

        Chiefly concerned with arguments over the acceptability
        of gravitational attraction as an unexplained property
        of matter.

373b. Eagles, Christina M., "David Gregory and Newtonian Science." *Brit. J. Hist. Sci.* 10 (1977): 216-225.

Examines the frequently made assertion that Gregory introduced Newtonianism into the universities of Scotland before it was taught at either Oxford or Cambridge, concluding on the basis of a large body of surviving manuscript evidence that his role has been greatly overstated. While Gregory himself responded enthusiastically to Newton's *Principia* and encouraged a few of his best students to study the work, he never made it an integral part of his lecture course while at Edinburgh.

374. Elkana, Yehuda, "Scientific and Metaphysical Problems: Euler and Kant." *Boston Studies in the Philosophy of Science, Vol. 14: Methodological and Historical Essays in the Natural and Social Sciences*, ed. R.S. Cohen and M.W. Wartofsky. Dordrecht/Boston: D. Reidel, 1974, pp. 277-305.

Discusses Euler's metaphysics, especially as it affected his theory of matter, and the influence of his outlook on Kant, which is claimed to be direct and important.

375. Frängsmyr, Tore, *Wolffianismens genombrott i Uppsala: Frihetstida universitetsfilosofi till 1700-talets mitt.* With English summary. Uppsala: Universitetsbiblioteket/ Stockholm: Almqvist and Wiksell, 1972. 258 pp.

Drawing largely on contemporary academic dissertations, traces the emergence of Wolffianism at Uppsala in the first half of the 18th century, led by the mathematicians Klingenstierna and Celsius. The controversy between Cartesian orthodoxy and Wolffian doctrines underlay most scientific debate at the time.

376. Frängsmyr, Tore, "Swedish Science in the 18th Century." *Hist. Sci.* 12 (1974): 29-42.

A brief outline of Swedish science c. 1700 to 1780, looking at the influence of economic, institutional, religious and philosophical factors. A good bibliography.

377. Freudenthal, Gad, "Littérature et sciences de la nature en France au début du XVIIIe siècle: Pierre Polinière, l'introduction de l'enseignement de la physique expérimentale à l'Université de Paris et l'*Arrêt burlesque* de Boileau." *Rev. d. synthèse* 101 (1980): 267-295.

Discusses the struggles over the introduction of the "new philosophy" into the University of Paris in the

years 1690-1710, emphasizing the strategic role played
in these by physics, which served as a Cartesian bridge-
head within the university. Draws attention to the
(modest) place in this story of Polinière's experimental
demonstrations, and also to his republishing at the
height of the battle of Boileau's famous satire, the
Arrêt burlesque, in a slightly modified form incorporating
a number of references to recent events.

378. Fueter, E., *Geschichte der exakten Wissenschaften in der
      Schweizerischen Aufklärung, 1680-1780*. Aarau/Leipzig:
      H.R. Säuerlander, 1941. xvi + 336 pp.

      Not seen.

379. Garin, Eugenio, "Antonio Genovesi e la sua introduzione
      storica agli *Elementa physicae* de Pietro van Musschen-
      broek." *Physis* 11 (1969): 211-222.

      Describes Genovesi's "ample" introduction to the 1745
      edition of Musschenbroek's *Elementa*, in which he stressed
      the importance of Galileo's work, criticized Descartes
      and examined Newton and Newtonianism extensively. Follows
      the considerable variations between editions as a repre-
      sentation of the introduction of Newtonianism in Italy.

380. Gay, John H., "Samuel Clarke on Matter and Freedom."
      J. Hist. Ideas 24 (1963): 85-105.

      Analyzes Clarke's refutation of various 17th-century
      viewpoints on the nature of matter and its relation to
      spirit, and discusses the failure of his own proposals,
      based on the assumptions of freedom and the correctness
      of Newton's ideas on matter and space. Shows that
      despite Clarke's use of the new "insights" of Newtonian
      philosophy, he was still attempting to solve basically
      17th-century problems.

381. Gibbs, F.W., "Itinerant Lecturers in Natural Philosophy."
      Ambix 8 (1960): 111-117.

      Provides brief biographical gleanings concerning the
      lecturers James Arden, Adam Walker and John Warltire,
      influenced by Priestley and active in the 1760s and
      1770s.

382. Gliozzi, Mario, *Fisici piemontesi del Settecento nel
      movimento filosofico del tempo*. Turin: Edizioni di
      "Filosofia," 1962. 16 pp.

Discusses the work of Giambatista Beccaria and the lively tradition of experimental physics that he inspired in Turin.

383. Gliozzi, Mario, "Lettres inédites de Giambattista Beccaria." *Proceedings, 11th International Congress of the History of Science, Warsaw, 1965* (Warsaw, 1968), vol. 3, pp. 210-215.

Publishes, with commentary, two letters of Beccaria's. One, to Paolo Frisi, which Gliozzi dates to shortly after the publication of Beccaria's *Dell'elettricismo artificiale e naturale* (1753), discusses the experimental basis of the Franklinian theory of electricity. The other, dated 20 July 1776, concerns inflammable air.

384. Gloden, Albert, "L'enseignement des sciences à l'ancien collège de Luxembourg au XVIIIe siècle." *Rev. Hist. Sci.* 11 (1958): 263-266.

Suggests that the college may have been valuable in the dissemination of *physique expérimentale*.

384a. Gowing, R., *Roger Cotes, Natural Philosopher*. Cambridge: Cambridge University Press, 1982. 200 pp.

Not seen.

* Gray, J.J., and Laura Tilling, "Johann Heinrich Lambert, Mathematician and Scientist, 1728-1777."

Cited herein as item 55.

385. Grigor'ian, A.T., "M.V. Lomonosov and his Physical Theories." *Arch. Int. Hist. Sci.* 16 (1963): 53-60.

A brief and rather uncritical account of Lomonosov's theories of heat, light and matter.

386. Guédon, Jean-Claude, "Chimie et matérialisme: la stratégie anti-newtonienne de Diderot." *Dix-huitième siècle* 11 (1979): 185-200.

Argues that Diderot rejected Newtonianism on account of the role it attributed to divine action in nature, advocating instead Rouelle's brand of chemistry which eschewed the possibility of knowledge about matter and restricted itself to a knowledge of processes.

387. Guenther, Siegmund, "Note sur Jean-André de Segner, fondateur de la météorologie mathématique." *Bolletino*

*di bibliografia e di storia delle scienze matematiche
e fisiche* 9 (1876): 217-28.

Brief description of the work of Segner (1704-77),
including his fluid mechanics, claiming that he was one
of the first German Newtonians.

388. Guerlac, Henry, "Newton's Changing Reputation in the 18th
      Century." *Carl Becker's Heavenly City Revisited*, ed.
      Raymond O. Rockwood. Ithaca, N.Y.: Cornell University
      Press, 1958, pp. 3-26. Reprinted, *Essays and Papers
      in the History of Modern Science* by Henry Guerlac.
      Baltimore/London: Johns Hopkins University Press,
      1977, pp. 69-81.

      Expounds the thesis, extensively developed by more
      recent writers, that central ideas of 18th-century
      natural philosophy such as the mechanistic world view
      and belief in the regularity of nature derived both from
      other works of Newton apart from the *Principia*, and from
      non-Newtonian sources.

389. Guerlac, Henry, "Where the Statue Stood: Divergent Loyal-
      ties to Newton in the 18th Century." *Aspects of the
      Eighteenth Century*, ed. Earl R. Wasserman. Baltimore:
      Johns Hopkins University Press, 1965, pp. 317-334.
      Reprinted, *Essays and Papers in the History of Modern
      Science* by Henry Guerlac. Baltimore/London: Johns
      Hopkins University Press, 1977, pp. 131-145.

      Suggests that Newton's reticence concerning the under-
      lying mechanisms of natural phenomena led both to his
      ideas being accepted only slowly on the Continent and
      to the appearance of widely varying, often contradictory,
      interpretations of his work.

389a. Guerlac, Henry, "The Newtonianism of Dortous de Mairan."
      *Essays on the Age of Enlightenment in Honor of Ira O.
      Wade*, ed. Jean Macary. Geneva/Paris: Librairie Droz,
      1977, pp. 131-141. Reprinted in slightly extended
      form, *Essays and Papers in the History of Modern
      Science* by Henry Guerlac. Baltimore/London: Johns
      Hopkins University Press, 1977, pp. 479-490.

      Argues that Mairan was by no means the "Cartesian of
      the strict observance" that he was depicted as by Brunet
      (cf. item 359), but that, on the contrary, he was one
      of the earliest in France to accept Newton's views on
      colour, and that he was also prepared, by the early
      1730s at the latest, to accept much of Newton's celestial
      physics.

390. Guerlac, Henry, *Essays and Papers in the History of Modern Science*. Baltimore/London: Johns Hopkins University Press, 1977. xx + 540 pp.

Contains reprints of a large number of the author's papers including "Newton's Changing Reputation in the 18th Century" (1958); "Francis Hauksbee: expérimentateur au profit de Newton" (1963); "Where the Statue Stood: Divergent Loyalties to Newton in the Eighteenth Century" (1965); and "The Newtonianism of Dortous de Mairan" (1976). Unfortunately the author's very important "Chemistry as a Branch of Physics: Laplace's Collaboration with Lavoisier" (item 612) is omitted from the collection; an earlier and much briefer paper on this subject is included instead.

391. Guerlac, Henry, "Some Areas for further Newtonian Studies." *Hist. Sci.* 17 (1979): 75-101. Reprinted, *Newton on the Continent* by Henry Guerlac. Ithaca/London: Cornell University Press, 1981, pp. 41-73.

Argues that there was not a sharp polarization between Cartesians and Newtonians in France at the beginning of the 18th century but that Malebranche and his followers occupied a mid-way position, and that in general there was too much appeal in Newton's work, in both the *Principia* and the *Opticks*, for it to be rejected outright by very many French natural philosophers, even those alleged to be "rigidly Cartesian."

392. Guerlac, Henry, *Newton on the Continent*. Ithaca/London: Cornell University Press, 1981. 169 pp.

Concerned mostly with the reception of Newton's ideas in France. Includes reprints of two previously published papers on the subject, plus a major new study (pp. 78-163), "Newton in France: The Delayed Acceptance of His Theory of Color." Also includes two short studies pointing to echoes of European thought in Newton's early notebook.

393. Hackmann, W.D., "The Growth of Science in the Netherlands in the 17th and Early 18th Centuries." *The Emergence of Science in Western Europe*, ed. M.P. Crosland. London: Macmillan, 1975, pp. 89-109.

Briefly describes the main influences on the growth of Dutch science: education, scientific institutions, religion, and popularization. Includes a good list of sources, mostly in Dutch.

394. Hahn, Roger, *Laplace as a Newtonian Scientist*. Los
      Angeles: William Andrews Clark Memorial Library,
      University of California, 1967.   21 pp.

      Draws various parallels between the career patterns
      and scientific work of Laplace and Newton but questions
      the usefulness of a label like "Newtonian" in describing
      any 18th-century scientist unless the term be further
      qualified. Briefly discusses Laplace's well-known
      determinist outlook, seeing this as a critical response
      to the views of d'Alembert and Condorcet.

394a. Hakfoort, C., "Christian Wolff tussen Cartesianen en
      Newtonianen." *Tsch. Gesch. Gnk. Natuurw. Wisk. Techn.*
      5 (1982): 27-38.

      Argues from an analysis of Wolff's views on light and
      on gravitation that the "polar view" which sees early
      18th-century science in terms of competing Cartesian and
      Newtonian schools is unsatisfactory, but that simply
      introducing another "-ism" such as neo-Cartesianism
      or Leibnizianism will not do.

395. Hall, A. Rupert, "Newton in France: A New View." *Hist.
      Sci.* 12 (1975): 233-250.

      Argues that the recognition and acceptance in France
      of Newton's mathematical and experimental work increased
      rapidly after 1715, "maturing" with the publication of
      a French translation of the *Opticks* in 1722 even though
      his mechanical and physical theories still had only
      limited acceptance at this time.

396. Hankins, Thomas L., *Jean d'Alembert: Science and the
      Enlightenment*. Oxford: Clarendon Press, 1970. xii +
      260 pp.

      The only adequate full length study of d'Alembert's
      science and its relation to the intellectual climate
      of the Enlightenment. Included are discussions of the
      effect of his rationalism on his rejection of forces,
      and analyses of the various controversies in which he
      was involved (especially with Euler and the Bernoullis)
      which helped to shape and define the science of rational
      mechanics. A good bibliography. Reviewed by R.H.
      Silliman, *Isis* 62 (1971): 255-257, R. Hahn, *Brit. J.
      Hist. Sci.* 66 (1972-73): 327-329, and E. Cane, "Jean
      d'Alembert between Descartes and Newton: A Critique of
      Thomas L. Hankins' Position." *Isis* 67 (1976): 274-276.

397. Hanna, Blake T., "Polinière and the Teaching of Experi-
      mental Physics at Paris, 1700-1730." *Eighteenth-
      Century Studies presented to Arthur M. Wilson*, ed.
      Peter Gay. New York: Russell and Russell, 1972, pp.
      13-39.

      Discusses Polinière's *Expériences de physique*, the
      standard text for experimental physics in the Jesuit
      colleges and the University until Nollet's work. In-
      cluded in Polinière's course were hydrostatics, capillary
      action, vacuum experiments, electroluminescence and
      optics.

398. Hardin, Clyde L., "The Scientific Work of the Reverend
      John Michell." *Ann. Sci.* 22 (1966): 27-47.

      Briefly describes the work of Michell (1724-1793).
      Discusses his ideas on artificial magnetism (including
      the first enunciation of the magnetic inverse square
      law), his development of a theory of matter similar to
      Boscovich's, and his independent development of the
      torsion balance to measure gravitational force.

399. Harrison, John Anthony, "Blind Henry Moyes, 'An Excellent
      Lecturer in Philosophy.'" *Ann. Sci.* 13 (1957): 109-
      125.

      Describes the life, travels and lectures of Moyes
      (1749-1807), an itinerant lecturer for twenty-nine
      years in Britain and the U.S.A. Provides summaries of
      some of his lecture topics including those on the
      "Galvanic Pile."

400. Heilbron, John L., *Electricity in the 17th and 18th
      Centuries: A Study of Early Modern Physics*. Berkeley/
      Los Angeles/London: University of California Press,
      1979. xiv + 606 pp.

      An outstanding and comprehensive history of electricity
      to about the year 1800, intended to illustrate the
      evolution of the experimental branches of physics more
      generally during the 17th and 18th centuries. To this
      end, the work also includes much information about
      changing theoretical conceptions during this period,
      and about the social and institutional setting in which
      the "electricians" worked. Essay review by R.W. Home,
      Ann. Sci. 38 (1981): 477-482. The first part of the
      work, together with a summary of the remainder, has
      been published separately as *Elements of Early Modern
      Physics* (Berkeley: University of California Press, 1982).

401. Heilbron, J.L., "Experimental Natural Philosophy." *The Ferment of Knowledge: Studies in the Historiography of Eighteenth-Century Science*, ed. G.S. Rousseau and Roy Porter. Cambridge: Cambridge University Press, 1980, pp. 357-387.

An excellent survey of the present state of research on 18th-century experimental physics, drawn substantially from the author's recent book (item 400) but extending the discussion of historiographical issues and offering a challenging list of desiderata for future work.

402. Heimann, P.M., "Newtonian Natural Philosophy and the Scientific Revolution." *Hist. Sci.* 11 (1973): 1-7.

Argues that it is an oversimplification to see all the many often mutually contradictory strands of 18th-century science as "Newtonianism," criticizing recent secondary literature for this. Stresses the necessity of studying the work of individual scientists with reference to their own particular religious, philosophical and intellectual backgrounds.

403. Heimann, P.M., "'Nature is a Perpetual Worker': Newton's Aether and 18th Century Natural Philosophy." *Ambix* 20 (1973): 1-25.

This important article traces the change from Newton's view that natural operations occurred through the action of God (manifested in active principles) on inert matter, via Boerhaave's notion of a fundamental active "material fire," to the later 18th-century view that active principles, in the form of innately active special forms of matter, were self-regulating parts of the natural order.

404. Heimann, P.M., "Voluntarism and Immanence: Conceptions of Nature in 18th-Century Thought." *J. Hist. Ideas* 39 (1978): 271-283.

Traces changing ideas of nature and matter from Newton, via Maclaurin and Hume, to Hutton and Priestley. Sees the Newtonian voluntarist conception of inactive matter controlled by divine will being rejected in favour of the idea that "activity was inherent in matter and immanent in Nature," with divine law being replaced by natural law.

405. Heimann, P.M., and J.E. McGuire, "Newtonian Forces and Lockean Powers: Concepts of Matter in 18th-Century Thought." *Hist. Studs. Phys. Sci.* 3 (1971): 233-306.

Argues that a number of 18th-century thinkers shared
a common conception that attractive and repulsive forces
were essential, inhering powers of matter. This notion
developed from the fundamentally different ideas of
Newton and Locke. Questions R.E. Schofield's thesis
(*Mechanism and Materialism*, item 456 herein) that there
were two fundamentally different matter theories prevalent
at this period. A very much shortened version of the
argument has been published by the same authors as "The
Rejection of Newton's Concept of Matter in the Eighteenth
Century," pp. 104-118 in Ernan McMullin, ed., *The Concept
of Matter in Modern Philosophy* (Notre Dame/London: Notre
Dame University Press, 1978).

*   Home, R.W., "'Newtonianism' and the Theory of the Magnet."

Cited herein as item 682.

406.   Home, R.W., "Out of a Newtonian Straightjacket: Alterna-
       tive Approaches to Eighteenth-Century Physical Science."
       Studies in the Eighteenth Century, IV, ed. R.F. Bris-
       senden and J.C. Eade. Canberra: Australian National
       University Press, 1979, pp. 235-249.

Argues that it can be not merely unhelpful but sometimes positively misleading to portray 18th-century
physical science in terms of a struggle between opposing
'Cartesian' and 'Newtonian' intellectual traditions.
Advocates instead an analysis based upon different 18th-
century attitudes towards the use of mathematics in
physics.

407.   Home, R.W., "Scientific Links between Britain and Russia
       in the Second Half of the Eighteenth Century." *Great
       Britain and Russia in the Eighteenth Century: Contacts
       and Comparisons*, ed. A.G. Cross. Newtonville, Mass.:
       Oriental Research Partners, 1979, pp. 212-224.

Presents evidence concerning, *inter alia*, the distribution in Britain of Aepinus' treatise on electricity and
magnetism, and the activities of J.H. de Magellan as a
"scientific agent" in late-18th-century London, especially his links with the St. Petersburg Academy of Sciences.

407a. Hoskin, Michael A., "'Mining All Within': Clarke's Notes
       to Rohault's *Traité de physique*." *Thomist* 24 (1962):
       353-363.

Details Samuel Clarke's increasingly Newtonian foot-
notes appended to successive editions of his Latin

translation of Rohault's enormously successful Cartesian-inspired treatise.

408. Hughes, Arthur, "Science in English Encyclopedias, 1704-1875. I." *Ann. Sci.* 7 (1951): 340-370.

Describes the publication and editorial history and the style of several English encyclopedias. Goes on to discuss the changing view of "final causes" and of "the Mechanical Philosophy," treated both from a religious viewpoint and with reference to criticisms of contemporaries.

409. Hughes, Arthur, "Science in English Encyclopedias, 1704-1875. II. Theories of the Elementary Composition of Matter." *Ann. Sci.* 8 (1952): 323-367.

Mainly concerned with theories concerning the elements, the first part of the 18th century showing much approval of Aristotle and Paracelsus. Includes a long section (pp. 347-367) on the treatment of "Fire" and "heat," showing the change from heat viewed as motion, to caloric matter, and back to motion.

409a. Hutchings, D.W., "Physical Science in Geneva during the Eighteenth Century (1740-1790)." M.Sc. thesis, University of London, 1960.

Not seen.

410. Iltis, Carolyn, "The Leibnizian-Newtonian Debates: Natural Philosophy and Social Psychology." *Brit. J. Hist. Sci.* 6 (1972-73): 341-377.

Argues that the Newtonians and Leibnizians of the 1720s were fundamentally opposed through a complex of metaphysical and mechanical ideas. Illustrates this by examining their totally different interpretations of the same experiments in the *vis viva* controversy.

411. Jacob, Margaret C., "Early Newtonianism." *Hist. Sci.* 12 (1974): 142-146.

Argues for the primary importance of the Boyle Lectures in first spreading Newton's ideas. The argument is developed at greater length in her *Newtonians and the English Revolution* (item 412).

412. Jacob, Margaret C., *The Newtonians and the English Revolution, 1689-1720*. Ithaca, N.Y.: Cornell University Press, 1976. 288 pp.

Argues that the triumph of Newtonian natural philosophy
in England during the period stated was facilitated, at
least, by its being adopted for polemical purposes by
the leaders of the dominant liberal wing of the Anglican
church. Reviewed by P.M. Heimann, *Hist. Sci.* 16 (1978):
143-151, and R.S. Westfall, *Amer. Hist. Rev.* 82 (1977):
353-355.

413. Jorgensen, Bent Soren, "Lomonosov, His Theory of Gravity
     and the Law of Conservation of Matter." *Physis* 17
     (1975): 21-40.

     Examines Lomonosov's Cartesian theory and others which
     assume a material cause of gravity. Argues that, far
     from Lomonosov being the discoverer of the law of con-
     servation of mass, his theory would actually have pre-
     vented him from conceiving the idea of testing a matter
     conservation law.

414. Kleinbaum, Abby R., "Jean Jacques Dortous de Mairan
     (1678-1771): A Study of an Enlightenment Scientist."
     Ph.D. dissertation, Columbia University, 1970. 257 pp.

     A general survey of Mairan's scientific ideas.

415. Kleinert, Andreas, *Die allgemeinverstandlichen Physik-
     bucher der französischen Aufklärung.* Aarau: Verlag
     Säuerlander, 1974. 187 pp.

     A very useful study of 18th-century French "popularisa-
     tions" of physics, and through them of the place of
     physics in the general French culture of the period.
     Concentrates on the first half of the century, and on
     the writings of Fontenelle, Algarotti, Regnault, Bou-
     geant, Voltaire and Pluche in particular.

416. Knight, Isabel F., *The Geometric Spirit: The Abbé de
     Condillac and the French Enlightenment.* New Haven/
     London: Yale University Press, 1968. xi + 320 pp.

     Describes the philosophy of Condillac (1914-1780), the
     chief French disciple of Locke and popularizer of New-
     ton, as the exemplar of the accepted ideas of science
     and methodology of his time. Includes a general dis-
     cussion of the science and mathematics of Condillac's
     time, and examines his rationalistic methodology which
     identified "science" with "geometry."

417. Kuhn, Albert J., "Glory or Gravity: Hutchinson vs. New-
     ton." *J. Hist. Ideas* 22 (1961): 303-322.

Describes the works of Hutchinson (1674-1737) and his influential Oxford-based followers, who, violently opposed to the mechanical philosophy, aimed to replace the Newtonian world-view by a Mosaic natural philosophy stressing revelation and derived by analogical arguments from etymological and typological exegesis of the Hebrew texts of the Bible.

418.   [Lambert, Johann Heinrich]. *Colloque international Jean-Henri Lambert (1728-1777)*. Paris: Editions Ophrys, 1979. 407 pp.

In addition to a number of papers on aspects of Lambert's life and his work in philosophy, mathematics and astronomy, includes the following on physical subjects: H. Pfeiffer, "Les études perspectives et chromatologiques de Lambert"; D. Speiser, "L'oeuvre de Lambert dans le domaine de l'optique"; R. Fox, "The Science of Fire: J.H. Lambert and Study of Heat"; P. Costabel, "Les 'Essais d'hygrométrie' de Lambert"; and L. Charbonneau, "Lambert et la physique mathématique du XIXe siècle."

419.   Lamontagne, Roland, "Lettres de Bouguer à Euler." *Rev. Hist. Sci.* 19 (1966): 225-246. Maheu, Gilles, "Introduction à la publication des lettres de Bouguer à Euler." *Rev. Hist. Sci.* 19 (1966): 206-224.

Maheu gives a brief biography of each of the correspondents and a short piece on the background to the various topics discussed by the two, including the principle of least action, refraction and electricity. Lamontagne's paper gives the texts of the letters themselves.

420.   Laudan, Laurens L., "Thomas Reid and the Newtonian Turn of British Methodological Thought." *The Methodological Heritage of Newton*, ed. Robert E. Butts and John W. Davis. Toronto: Toronto University Press/Oxford: Basil Blackwell, 1970, pp. 103-131.

Discusses Reid's philosophy of science which blended the empiricist and inductivist traditions, arguing that Reid was the first of many British philosophers to take Newton's epistemology and methodology seriously.

421.   Levere, T.H., "Relations and Rivalry: Interactions between Britain and the Netherlands in 18th-century Science and Technology." *Hist. Sci.* 9 (1970): 42-53.

Outlines largely unexplored areas of scientific interaction between the two economically and politically

linked countries, especially the rapid transference of
Newtonianism to Holland, the spread of English "popular"
science, the large numbers of British students in Holland
and the close links between the two countries' instrument makers and experimental physicists. A good select
bibliography of primary and secondary work.

422. Levere, T.H., "Friendship and Influence: Martinus van
Marum, F.R.S." *Notes Rec. Roy. Soc. Lond.* 25 (1970):
113-120.

Describes van Marum's active lobbying to become F.R.S.
from c. 1787 until his eventual success in 1798, showing
the networks of scientific contacts existing between
England and Holland in the 18th century.

422a. Loria, Mario, "Un manuscrit de l'Académie des Sciences
de Turin: le *Traité de physique* de Jean-Baptiste
Beccaria (1754)." *Proceedings, 12th International
Congress of the History of Science, Paris, 1968*
(Paris, 1971), Vol. 3B, pp. 71-76.

Briefly describes Beccaria's unpublished treatise,
focusing especially on the discussion of electricity
therein.

423. McCormmach, Russell, "John Michell and Henry Cavendish: Weighing the Stars." *Brit. J. Hist. Sci.* 4
(1968): 126-155.

Argues that Michell and Cavendish did not follow the
strongly empiricist methods of their contemporaries,
but sought fundamental mathematical laws of attractive
and repulsive particles, in this case between light and
the matter of the stars.

424. McCormmach, Russell K., "Henry Cavendish: A Study of
Rational Empiricism in 18th-Century Natural Philosophy."
Isis 60 (1969): 293-306.

Argues from an examination of different aspects of
Cavendish's work that his varied researches were in
fact "integral parts of a persistent scientific goal"
based on the corpuscular and dynamical ideas of Newton's *Opticks* and *Principia*.

424a. Machabey, Armand, "Vue sommaire sur quelques rapports
entre l'*Encyclopédie* et la métrologie." *L'"Encyclopédie" et le progrès des sciences et des techniques*,
ed. S. Delorme and R. Taton. Paris: Presses Univer-
sitaires de France, 1952, pp. 215-224.

Focuses on the preoccupation, in the *Encyclopédie*
articles on mensuration, on the need to establish standards and on the value of a decimal division of measures.
Also discerns in these articles a clearer perception
than previously of the notion of error.

425. McKie, Douglas, "Priestley's Laboratory and Library and
    Other of his Effects." *Notes Rec. Roy. Soc. Lond.* 12
    (1956-57): 114-136.

    Publishes Priestley's lists of the apparatus and
    scientific books destroyed in the Birmingham Riots,
    which he presented in claiming for damages to the King's
    Bench. The lists include a large number of electrical
    and optical instruments.

426. Maddison, R.E.W., and Francis R. Maddison, "Joseph
    Priestley and the Birmingham Riots." *Notes Rec. Roy.
    Soc. Lond.* 12 (1956-57): 98-113.

    Describes the circumstances of the riots and the
    pecuniary losses Priestley sustained.

427. Marsak, Leonard M., "Bernard de Fontenelle: The Idea of
    Science in the French Enlightenment." *Trans. Amer.
    Phil. Soc.* 49 (7) (1959): 64 pp.

    Portrays Fontenelle as a serious critic and philosopher
    of science, and includes a discussion of his Cartesianism
    and his philosophy of science.

428. Marsak, Leonard M., "Cartesianism in Fontenelle and
    French Science, 1686-1752." *Isis* 50 (1959): 51-60.

    Argues that Fontenelle was Cartesian only in his ac-
    ceptance of Descartes' cosmology and skeptical method.
    In his rejection of a rationalist methodology and meta-
    physics and his espousal of a critical empiricism he was
    anti-Cartesian.

429. Mautner, Franz H., and Franklin Miller, Jr., "Remarks on
    G.C. Lichtenberg, Humanist-Scientist." *Isis* 43 (1952):
    223-231.

    Describes some of Lichtenberg's speculations on science
    and also his methodology, including his espousal of the
    unity of the sciences, the rejection of the "wild guess"
    from science, and the distinction between reality and a
    scientific theory agreeing with observation. This last
    allowed him to adopt an agnostic position in debates

such as that between the fluid theories of electricity and on the nature of light.

430. Metzger, Hélène, *Newton, Stahl, Boerhaave et la doctrine chimique*. Paris: A. Blanchard, 1930. 332 pp.

Despite its age, still a valuable reference. Comprises three sections: the first, on Newton and his influence on 18th-century matter theory, including his views on the influence of light on matter in chemical reactions; the second on Stahl and his phlogiston and matter theories; the last on Boerhaave including an extensive discussion of his influential theory of elemental fire.

431. Metzger, Hélène, *Attraction universelle et religion naturelle chez quelques commentateurs anglais de Newton*. Paris: Hermann, 1938. 223 pp. (Actualités scientifiques et industrielles, 621-623).

A classic study of the theological implications drawn from Newton's science by a variety of 18th-century English commentators.

432. Monchamp, Georges, *Histoire du Cartésianisme en Belgique*. Brussels: F. Hayez, 1886. 643 pp.

The last two chapters provide brief accounts of the work of various 18th-century Cartesian philosophers in Belgium, at a time when Cartesianism had become accepted and orthodox.

433. Moüy, Paul, *Le développement de la physique cartésienne 1646-1712*. Paris, 1934. Reprinted, New York: Arno Press, 1981.

A classic study. Mostly concerned with 17th-century developments but also includes an important discussion of Malebranche's physics.

433a. Musson, A.E., and Eric Robinson, *Science and Technology in the Industrial Revolution*. Manchester: Manchester University Press, 1969. viii + 534 pp.

The authors argue, on the basis of detailed study of some selected themes drawn chiefly from the engineering and chemical industries, that developments in science and in technology in Britain during the 18th century were not unrelated, and that "the Industrial Revolution was also an intellectual movement." Includes much valuable material on the diffusion of science and technology in 18th-century Britain and also on the work of James Watt.

434. Nedelkovich, Dusan, *La philosophie naturelle et relati-*
      viste de R.J. Boscovich. Paris: Editions de la Vie
      Universitaire, 1923. 241 pp.

      One of the best discussions of Boscovich's theory of
      matter and law of continuity. Also contains a useful
      discussion of Boscovich's philosophy of space and time
      and briefly compares his philosophy with that of Locke,
      Newton, Leibniz and Kant.

435. Olson, Richard, "The Reception of Boscovich's Ideas in
      Scotland." *Isis* 60 (1969): 91-103.

      Argues that an important line of transmission at the
      end of the 18th century for Boscovich's ideas of matter
      to Britain was through the Scottish Common Sense
      philosophers--especially Dugald Stewart--and scientists
      trained in that tradition such as Robison, Playfair and
      Leslie.

436. Parry, J.S., "John Michell's Theory of Matter and
      Joseph Priestley's Use of It." M. Phil. thesis,
      University of London, 1977. 231 pp.

      An extended discussion of Michell's ideas, linking
      them to his early work on magnetism and emphasizing
      differences between Michell's conceptions and those of
      Boscovich. Shows how Priestley later adapted Michell's
      views in developing his own ideas on matter and spirit.

437. Pater, Cornelis de, *Petrus van Musschenbroek (1692-1761),*
      een newtoniaans natuuronderzoeker. With a summary in
      English. Utrecht: Drukkerij Elinkwijk BV, 1979. x +
      382 pp.

      The only full-length monograph on Musschenbroek.
      Examines the Newtonianism of his methodology in com-
      parison with that of his contemporaries, especially
      'sGravesande. After briefly discussing his biography and
      the intellectual background, de Pater characterizes
      Musschenbroek's Newtonianism as having a strong empirical
      basis with a Baconian emphasis on the accumulation of
      pure data, coupled with a rejection of *a priori* reason-
      ing and a violent anti-Cartesianism. Gives detailed
      analyses of Musschenbroek's now neglected work on mag-
      netism and capillary action, emphasizing his application
      of (or deviations from) his stated methodology, his use
      of attractive and repulsive forces and his rejection of
      vortical and effluvial models. Concludes by slotting
      Musschenbroek into the standard "physico-theology"
      tradition of his time. An excellent Musschenbroek bib-
      liography, and a good secondary bibliography.

438. Pater, Cornelis de, "Petrus van Musschenbroek (1692–1761): A Dutch Newtonian." *Janus* 64 (1977): 77–87.

A brief account, in English, based upon the author's Dutch thesis on Musschenbroek (item 437). Emphasizes Musschenbroek's empiricism, contrasting his approach somewhat to that of his contemporary, 'sGravesande.

439. Philip, J.R., "Samuel Johnson as Antiscientist." *Notes Rec. Roy. Soc. Lond.* 29 (1974–75): 193–203.

Shows that even though Johnson respected the "scientific method," he had a strong dislike for science on account of what he saw as its trivial nature and dehumanizing effects.

440. Pighetti, C., "Discorrendo del newtonianesimo di R.G. Boscovich." *Physis* 6 (1964): 15–27.

Looks at the Newtonian elements of Boscovich's theories of space, matter, force, and so on.

441. Quetelet, L.A.J., *Histoire des sciences mathématiques et physiques chez les Belges*. Brussels: M. Hayez, 1864. 479 pp.

Links the often troubled political history of Belgium with its science up to the early 19th century, with an emphasis on mathematics. Notes a general decline in all aspects of Belgian science in the 18th century. Contains useful, roughly chronological, listings of practitioners and their works.

442. Quinn, Arthur J., "Evaporation and Repulsion: A Study of English Corpuscular Philosophy from Newton to Franklin." Ph.D. dissertation, Princeton University, 1970.

Sees evaporation as a central problem in the Newtonian philosophy because the corpuscular repulsion it seemed to entail needed to be unified with more general aspects of Newtonianism. Specifically examines Desaguliers' hypothesis of evaporation by electrification, and Franklin's failed attempt to unify what he saw as two mechanisms of evaporation, namely that by heat repulsion and that by "another" mechanism.

442a. Quinn, Arthur, "Repulsive Force in England, 1706–1744." *Hist. Studs. Phys. Sci.* 13 (1982): 109–128.

Surveys British attitudes, during the period stated, to Newton's proposal in the 31st Query in his *Opticks* that nature employed repulsive as well as attractive forces between particles. Argues that there was a significant pattern of development from Freind in the early

years of the century through Hales to Desaguliers in
the early 1740s.

443. Ravetz, J.R., "The Representation of Physical Quantities
      in 18th-Century Mathematical Physics." *Isis* 52
      (1961): 7-20.

      Describes the means by which Euler, Atwood and Lagrange
      approached the problem of dimensionality of physical
      quantities.

443a. Robinet, André, *Malebranche de l'Académie des Sciences:*
      l'oeuvre scientifique, 1674-1715. Paris: J. Vrin,
      1970. 447 pp.

      Chiefly concerned with Malebranche's work in the last
      decades of the 17th century in mathematics and in rec-
      tifying the Cartesian principles of mechanics. Also,
      however, includes a substantial discussion of his later
      work in optics, both in developing the undulatory
      theory of light and in helping to win acceptance in
      France for Newton's theory of colour.

444. Robinson, Eric, and Douglas McKie, eds., *Partners in*
      Science: Letters of James Watt and Joseph Black.
      London: Constable, 1970. xvi + 502 pp.

      A fascinating and important collection that also in-
      cludes a large number of letters that passed between
      Watt and John Robison, and sundry others that bear upon
      the main themes of the correspondence.

445. Robinson, Eric, "Priestley's Library of Scientific
      Books: A New List." *Studs. Hist. Phil. Sci.* 1 (1979):
      145-160.

      On the basis of previously published lists and a new
      manuscript, constructs a 'minimum' list of books owned
      by Priestley and destroyed in the Birmingham riots of
      1791. Shows thereby that Priestley had far wider and
      more varied sources of information available to him,
      both British and international, than previously believed.

446. Roche, Daniel, "Science et pouvoirs dans la France du
      XVIIIe siècle, 1666-1803." *Annales* 29 (1974): 738-748.

      Essay review of Roger Hahn's *The Anatomy of a Scien-*
      tific Institution: The Paris Academy of Sciences, 1666-
      1803 (1971) (item 485).

447. Rousseau, G.S., and Roy Porter, eds., *The Ferment of*
      Knowledge: Studies in the Historiography of Eighteenth-
      Century Science. Cambridge: Cambridge University
      Press, 1980. xiv + 500 pp.

A collection of essays intended "to survey, synthesize, appraise and criticize new trends and major controversies" in research on 18th-century science. Papers relating to physics are noted individually elsewhere in this bibliography. Essay review by G.N. Cantor, *Hist. Sci.* 20 (1982): 44-63.

448. Rowbottom, Margaret E., "The Teaching of Experimental Philosophy in England, 1700-30." *Actes du XIe Congrès International d'Histoire des Sciences, Warsaw, 1965* (Warsaw, 1968), Vol. 4, pp. 46-53.

A useful discussion of the rise of the major lecturers and teachers in London and Oxford, including Cotes, Hauksbee, Whiston, Desaguliers and Keill.

448a. Rowbottom, Margaret E., "John Theophilus Desaguliers (1683-1744)." *Proc. Huguenot Soc. Lond.* 21 (1965-70): 196-218.

Includes many valuable details concerning courses of experimental lecture-demonstrations in early 18th-century England, together with a useful survey of Desaguliers' scientific work.

449. Ruestow, Edward G., *Physics at 17th and 18th-Century Leiden.* (Archives internationales d'histoire des idées, series minor, no. 11.) The Hague: Martinus Nijhoff, 1973. 174 pp.

Discusses the rise of physics at Leiden. Concentrates on the 17th century, but has a chapter on the Newtonianism of 'sGravesande and Musschenbroek and their influence, teaching, philosophy and experiments. A conclusion deals with the pedagogical role of the physicists, and reasons why the University was unable to maintain its lead in physics in the latter half of the 18th century. Valuable bibliography on Dutch education and universities in the 16th-18th centuries.

450. Sadoun-Goupil, Michelle, "La correspondance de Claude-Louis Berthollet et Martinus Van Marum (1786-1805)." *Rev. Hist. Sci.* 25 (1972): 221-252.

Introductory comments followed by a list of the 33 letters between the two scientists preserved in the archives of the Hollandsche Maatschappij der Wetenschappen, with a summary of their primarily scientific contents. Includes several items on physics, especially electricity.

450a. Sarton, George, "The Study of Early Scientific Text-
       books." *Isis* 38 (1947-48): 137-148.

       Calls for the study not just of the main discoveries
       of science, but also of "how these discoveries were re-
       ceived by the people and transmitted by them to their
       neighbours or to the coming generation." Illustrates
       the value of this approach by discussing the various
       editions of Rohault's famous *Traité de physique*. Accom-
       panied (pp. 149-150) by "Comments" by I. Bernard Cohen.

451.   Schaffer, Simon, "Natural Philosophy." *The Ferment of
       Knowledge: Studies in the Historiography of Eighteenth-
       Century Science*, ed. G.S. Rousseau and Roy Porter.
       Cambridge: Cambridge University Press, 1980, pp. 55-
       91.

       A critical review of recent historical literature in
       which he argues against the thesis that 18th-century
       natural philosophy was essentially in a Newtonian
       tradition, proposing instead an alternative attitude
       incorporating the methodological ideas of Bachelard,
       Kuhn and Foucault. Extensive secondary bibliography.

452.   Schimank, Hans, "Die Wandlung des Begriffs 'Physik'
       während der ersten Hälfte des 18. Jahrhunderts."
       *Wissenschaft, Wirtschaft und Technik: Studien zur
       Geschichte*, ed. Karl-Heinz Manegold. Munich: Bruck-
       mann, 1969, pp. 454-468.

       Argues on the basis of a number of leading 18th-
       century physics textbooks that the general conception of
       the subject changed significantly during the period,
       from the old-style general "philosophy of nature" to
       an experimentally based body of knowledge.

453.   Schofield, Robert E., "John Wesley and Science in 18th-
       Century England." *Isis* 44 (1953): 331-340.

       Describes Wesley's publications on science, which
       were rather haphazard syntheses from a large variety
       of sources. Wesley recommended empirical science as
       a diversion, as a means of appreciating God's work, and
       for its usefulness, but rejected theory utterly (especial-
       ly Newtonianism and mathematics) as a futile endeavour
       to understand God's ways.

454.   Schofield, Robert E., "The Scientific Background of
       Joseph Priestley." *Ann. Sci.* 13 (1957): 148-163.

Argues that Priestley was far from being the untutored and self-trained scientist he himself claimed to be in his *Memoirs*.

455. Schofield, Robert E., "Joseph Priestley: The Theory of Oxidation and the Nature of Matter." *J. Hist. Ideas* 25 (1964): 285-294.

Argues for the central importance of Boscovichean atomism in Priestley's science and theology.

456. Schofield, Robert E., *Mechanism and Materialism: British Natural Philosophy in an Age of Reason.* Princeton, N.J.: Princeton University Press, 1970. viii + 336 pp.

An extended account of Newton's legacy in 18th-century British natural philosophy, arguing that this manifested itself in two forms: (1) an initial "mechanist" phase in which all of nature was held to be reducible to forces acting at a distance between material corpuscles, and (2) a later "materialist" phase in which subtle fluids were invoked as mediating agencies in bringing about natural effects. A final section argues for a third phase of "neo-mechanism" in the period after 1760 as manifested in the writings of men such as Michell, Cavendish, William Herschel, Priestley and Hutton. Essay review by Yehuda Elkana, *Brit. J. Phil. Sci.* 22 (1971): 297-306.

457. Schofield, Robert E., "An Evolutionary Taxonomy of Eighteenth-Century Newtonianisms." *Studies in Eighteenth-Century Culture* 7 (1978): 175-192.

Surveys the diffusion and diversification of Newton's ideas across 18th-century Europe, delineating in the process such an astonishing variety of "Newtonianisms" that one is left doubting whether the term has any real signification at all.

458. Shapin, Steven, "Social Uses of Science." *The Ferment of Knowledge: Studies in the Historiography of Eighteenth Century Science*, ed. G.S. Rousseau and Roy Porter. Cambridge: Cambridge University Press, 1980, pp. 93-139.

A critical examination of the bases of recent studies investigating the interaction of science with social factors. Includes discussions on historical writings concerned with Newtonianism and Enlightenment matter-theory. Good bibliography.

459.   Sheynin, O.B., "Origin of the Theory of Errors." *Nature*
       211 (1966): 1003-1004.

       Argues that Lambert should be given precedence over
       Gauss as the originator of the theory of errors, and
       describes briefly a number of Lambert's important con-
       tributions to the subject.

459a. Smolka, Josef, "L'abbé Nollet et la physique en Bohême."
       *Proceedings, 12th International Congress of the History
       of Science, Paris, 1968* (Paris, 1971), Vol. 3B, pp.
       131-135.

       Details Nollet's influence on mid-18th-century Bohemian
       experimental physicists as illustrating his importance
       in the diffusion of science at this period.

460.   Stewart, Larry, "Samuel Clarke, Newtonianism, and the
       Factions of Post-Revolutionary England: Anti-Newtonianism
       in England, 1689-1720." *J. Hist. Ideas* 42 (1981):
       53-72.

       Sees the High Church Tories, especially the anti-
       Newtonian Robert North, as being in opposition to the
       "faction" of Whiggish, Newtonian, Low Churchmen on
       grounds of science, metaphysics and politics.

461.   Strong, Edward W., "Newtonian Explications of Natural
       Philosophy." *J. Hist. Ideas* 18 (1957): 49-83.

       Argues that Newton distinguished in his methodology
       between knowledge gained by mathematical reasoning and
       that gained by observation. Of the major British New-
       tonians, only Keill was a "mathematical realist," be-
       lieving in the reification of mathematical concepts.
       The others--Pemberton, 'sGravesande, Maclaurin, Whiston
       and Desaguliers--were 'mathematical conceptualists'
       who distinguished between physical reality and mathe-
       matical formulations.

462.   Taton, René, "A propos de l'oeuvre de Monge en physique."
       Rev. Hist. Sci. 3 (1950): 174-179.

       Publishes three little-known notes of Monge on physics,
       including one on acoustic theory and another on caloric.

463.   Taton, René, ed., *Enseignement et diffusion des sciences
       en France au XVIIIe siècle.* Paris: Hermann, 1964.
       780 pp.

The common theme underlying the twenty-three papers presented here is to examine the diffusion and teaching of scientific ideas through the various educational establishments of 18th-century France, linking them to the general trends of education and intellectual life. Papers of particular interest to our field include Roger Hahn on the *Ecoles Militaires* and the *Observatoires*; de Dainville, Costabel, Lemoine, Lacoarret and Ter-Menassian on the *Collèges* of the religious orders and the universities; Jean Torlais on *physique expérimentale*; and René Taton on the *Ecole du Génie* at Mézières.

464. Taton, René, "Madame du Châtelet, traductrice de Newton." *Arch. Int. Hist. Sci.* 22 (1969): 185-210.

While providing some details of du Châtelet's earlier scientific work, Taton concentrates on the period when she was translating the *Principia*, describing the origin of the project and the background of discussions with Clairaut, Algarotti, Jean Bernoulli and others, concluding with an account of the posthumous publication of the work.

465. Thackray, Arnold, "'Matter in a Nut Shell': Newton's *Opticks* and Eighteenth-Century Chemistry." *Ambix* 15 (1968): 29-53.

Describes Newton's theory, as set out in his *Opticks*, that matter was composed mainly of empty space. Discusses the rapid adoption of the idea by writers such as the Keills, Pemberton, Voltaire and Boerhaave, and its eventual development into the point force atomism of Boscovich and Priestley.

465a. Thackray, Arnold, "'The Business of Experimental Philosophy': The Early Newtonian Group at the Royal Society." *Proceedings, 12th International Congress on the History of Science, Paris, 1968* (Paris, 1971), Vol. 3B, pp. 155-159.

Suggests that the establishment of Newtonianism as scientific orthodoxy in early 18th-century England involved "such social factors as political exigency, national pride, group interest and private ambition" as well as the more frequently discussed intellectual and priority considerations.

466. Thackray, Arnold, *Atoms and Powers: An Essay on Newtonian Matter-Theory and the Development of Chemistry.* Cam-

bridge, Mass.: Harvard University Press/London: Oxford University Press, 1970. xxvi + 326 pp.

A study of the transformations that Newton's theory of matter underwent at the hands of his 18th-century followers, culminating in an important discussion of the origins of Dalton's atomic theory. Reviewed by Roger Hahn, *Isis* 62 (1971): 554-555, and Yehuda Elkana, *Brit. J. Phil. Sci.* 22 (1971): 297-306.

467. Turner, Gerard L'E., "The Portuguese Agent: J.H. de Magellan." *Antiquarian Horology* 9 (1974): 74-77.

A brief description of the life of Magellan (1722-1790), who acted as a London-based agent for purchasers of instruments and books as well as a disseminator of information on instruments and scientific work including, most notably, Adair Crawford's theory of heat.

468. Turner, G.L'E., "Henry Baker, F.R.S., Founder of the Bakerian Lecture." *Notes Rec. Roy. Soc. Lond.* 29 (1974-75): 53-79. Reprinted, *Essays on the History of the Microscope* by G.L'E. Turner. Oxford: Senecio Publishing Co., 1980, pp. 185-213.

Shows Baker to have been a talented dilettante whose scientific interest led him to publish a successful popularization of microscopy (1742) and a study on crystals in solution (1753), to make a valuable contribution to microscope design, and to dabble in electrical studies (1745-1753).

468a. Vanpaemel, G., "Cartesiaanse en Newtoniaanse natuur-wetenschap aan de Leuvense Artesfakulteit." *Tsch. Gesch. Gnk. Natuurw. Wisk. Techn.* 5 (1982): 39-49.

Analyzes the dissemination of Cartesian and Newtonian science into the Arts curriculum at the University of Louvain, arguing that in this case there was no sharp controversy between the two systems, and that it is difficult to distinguish clearly between "Cartesian" and "Newtonian" points of view in the curriculum.

469. Whyte, Lancelot L., "R.J. Boscovich, S.J., F.R.S. (1711-1787), and the Mathematics of Atomism." *Notes Rec. Roy. Soc. Lond.* 13 (1958): 34-48.

Provides a brief biography and a concise account of Boscovich's atomic theory, and suggests several scientists of the 18th and 19th centuries who may have been influenced by him.

470. Wightman, W.P.D., "The Copley Medal and Some of its Early Recipients." *Physis* 3 (1961): 344-355.

    The first medal was awarded in 1736 and the first twenty recipients included Franklin, William Watson, Benjamin Robins and Desaguliers.

471. Wilde, C.B., "Hutchinsonianism, Natural Philosophy and Religious Controversy in Eighteenth Century Britain." *Hist. Sci.* 18 (1980): 1-24.

    Presents the ideas of Hutchinson and his followers in some detail as representing a persisting strand of anti-Newtonian thought in 18th-century Britain.

472. Wilde, C.B., "Matter and Spirit as Natural Symbols in Eighteenth-Century British Natural Philosophy." *Brit. J. Hist. Sci.* 15 (1982): 99-131.

    Explores some of the controversies that took place in 18th-century Britain over the concepts of matter and spirit, viewing these from an anthropological perspective drawn from the work of Mary Douglas.

472a. Willemse, David, "Suites d'un voyage aux Pays-Bas: João Jacinto de Magalhães (1722-1790) et ses rapports avec Jean Henri Van Swinden (1746-1823)." *Arquivos do Centro Cultural Português* 7 (1974): 225-278.

    Publishes the text of 17 letters exchanged between Magellan and Van Swinden between 1777 and 1786, together with an introduction.

473. Wolf, A., *A History of Science, Technology, and Philosophy in the Eighteenth Century.* London: George Allen & Unwin, 1938. 814 pp.

    An encyclopedic work, much the best of its kind for the period in question. Includes a chapter on 18th-century developments in mechanics, four on the different branches of physics, and useful chapters as well on astronomical, marine and meteorological instruments and the making of lenses and specula.

474. Zemplén, J., "The Cartesianism in the Physics of Hungary." *Proceedings, 13th International Congress of the History of Science, Moscow, 1971* (Moscow, 1974), Vol. 6, pp. 226-232.

    Argues that, despite some knowledge of Newton's physics, Cartesian attitudes remained strong in Hungarian

physics textbooks well into the second half of the 18th
century.

(b) Institutional Histories

475. Barrière, P., *L'Académie de Bordeaux: centre de culture
      internationale au XVIIIe siècle (1712-1792)*. Bordeaux/
      Paris: Editions Bière, 1951. xii + 374 pp.

      A very thorough account of the history of this impor-
      tant provincial academy, including an extended discussion
      of the place of experimental physics in the Academy's
      activities.

*   Brockliss, L.W.B., "Aristotle, Descartes and the New
        Science: Natural Philosophy at the University of
        Paris, 1600-1740."

        Cited herein as item 357.

476. Cantor, G.N., "The Academy of Physics at Edinburgh,
      1797-1800." *Soc. Studs. Sci.* 5 (1975): 109-134.

      Describes the membership and activities of this short-
      lived society, seen as being formed in reaction against
      the Tory-dominated Royal Society of Edinburgh by a
      group of middle-class, Whig-orientated younger men.

*   Christie, J.R.R., "The Origins and Development of the
        Scottish Scientific Community, 1680-1760."

        Cited herein as item 367.

*   Christie, J.R.R., "The Rise and Fall of Scottish Science."

        Cited herein as item 368.

477. Cochrane, Rexmond C., "Francis Bacon and the Rise of
      the Mechanical Arts in 18th-Century England." *Ann.
      Sci.* 12 (1956): 137-156.

      Argues for a widespread dissemination of mathematical
      and physical ideas during the century due to the multi-
      tude of texts and lecture courses available and an in-
      creasing and overtly Baconian stress on utility that
      led to the foundation of various societies which actively
      encouraged technical education and innovation.

478. Cousin, Jean, "L'Académie des sciences, belles lettres et arts de Besançon au XVIIIe siècle et son oeuvre scientifique." *Rev. Hist. Sci.* 12 (1959): 327-344.

     Includes a description of the physical apparatus at the Besançon Academy and the experiments conducted there in imitation of the work of leading contemporary electricians.

479. Dainville, François de, "L'enseignement des mathématiques dans les collèges Jésuites de France du XVIe au XVIIIe siècle." *Rev. Hist. Sci.* 7 (1954): 6-21, 109-123.

     Looks at the teachers, the distribution of the colleges and some of the work done in them. *Mathématiques* included optics, mechanics and hydraulics, and astronomy.

480. Dulieu, Louis, "Le mouvement scientifique montpellierain au XVIIIe siècle." *Rev. Hist. Sci.* 11 (1958): 227-249.

     Recounts the activities, including teaching and research, of those interested in science in Montpellier, including the holders of various posts in the University and the local Academy. Gives some information, in particular, concerning Bertholon's occupancy of a chair of physics in the Academy, 1781-92. Includes a useful bibliography.

481. Dulieu, Louis, "La contribution montpellieraine aux recueils de l'Académie Royale des Sciences." *Rev. Hist. Sci.* 11 (1958): 250-260.

     Argues that the 18th-century Montpellier academicians were the most serious of the French provincial scientists, and lists their publications in the *Histoire de l'Académie Royale des Sciences de Paris*. Only four papers on physics.

482. Duveen, Denis I., and Roger Hahn, "Laplace's Succession to Bézout's Post of 'Examinateur des Elèves de l'Artillerie': A Case History in the 'Lobbying' for Scientific Appointments in France during the Period preceding the French Revolution." *Isis* 48 (1957): 416-427.

     First describes the limited ways by which a scientist could earn a livelihood in 18th-century France, and then uses a letter from Laplace to Lavoisier to illustrate the manner in which appointments to these positions

were made, with the candidates relying upon the rapid
and unabashed mobilization of friends and colleagues in
positions of influence.

483. Fay, Bernard, "Learned Societies in Europe and America
      in the 18th Century." *Amer. Hist. Rev.* 37 (1932):
      255-266.

      Briefly identifies the different types of learned
      societies that flourished in the period and their dis-
      tribution. Conjectures that their influence was pri-
      marily social rather than productive of research.

484. Hahn, Roger, "The Chair of Hydrodynamics in Paris, 1775-
      1791: A Creation of Turgot." *Actes du Xe congrès
      international d'histoire des sciences, Ithaca, 1962*
      (Paris, 1964), vol. 2, pp. 751-754.

      This chair, occupied for most of its history by
      Charles Bossut, is presented not as an important
      development for the science of hydrodynamics--that it
      was of any importance in this regard is expressly
      denied--but as an illustration of an *ancien régime*
      scientific institution.

485. Hahn, Roger, *The Anatomy of a Scientific Institution:
      The Paris Academy of Sciences, 1666-1803.* Berkeley,
      Los Angeles and London: University of California
      Press, 1971. xiv + 433 pp.

      An outstanding study which traces the rise of the
      Academy from its origins in the salons of 17th-century
      Paris, its evolution from a semi-private organization
      into the arbiter of scientific standards in *ancien
      régime* France, through the turmoil of the Revolution,
      the disbanding of the organization and its reincarnation
      as the fundamentally different Institut de France.
      Since the author is concerned with the Academy as an
      organic unit, he does not detail the work of individuals
      except in so far as this reflects the changing role of
      the Academy. Includes an excellent bibliography. Re-
      viewed by M.P. Crosland, *Isis* 63 (1972): 405-407; K.M.
      Baker, *Minerva* 10 (1972): 502-508; H. Brown, *Ann. Sci.*
      29 (1972): 313-316.

486. Hahn, Roger, "Scientific Careers in Eighteenth-Century
      France." *The Emergence of Science in Western Europe*,
      ed. M.P. Crosland. London: Macmillan, 1975, pp. 127-
      138. Also published as "Scientific Research as an

Occupation in Eighteenth-Century Paris." *Minerva* 13 (1975): 501-513.

Argues that while a class of professional scientific practitioners emerged in 18th-century France, for economic, social and practical reasons these men still spent the bulk of their time on non-research activities. Includes a good bibliography.

487. Kopelevich, Yu. Kh., *Osnovanie Peterburgskoi akademii nauk*. (The Formation of the Petersburg Academy of Sciences). Leningrad: Nauka, 1977. 212 pp.

An excellent history of the institutional aspects of the early years of the Academy (1725-1747) which specifically does not concern itself with the scientific work of the academicians. The first half of the book traces the institution's conception and establishment, and includes an account of the means by which prominent European scientists were enlisted to establish an advanced institute in scientifically backward Russia. The second half examines the Academy's organizational evolution, the question of its purpose and its relationship with the government and other scientific bodies. The author liberally uses both archival and published sources, and ensures a clear understanding of the historical context. Reviewed by Nathan M. Brooks, *Isis* 72 (1981): 107-108.

488. Kronick, D.A., "Scientific Journal Publication in the 18th Century." *Bibliographical Society of America*. *Papers* 59 (1965): 28-44.

Drawing on very scattered information, describes the frequent non-profitability of such ventures and their high failure rate, examines some "official" and private sponsorships, estimates some (highly variable) print runs and argues for generally poor distribution.

489. Layton, David, "Diction and Dictionaries in the Diffusion of Scientific Knowledge: An Aspect of the History of the Popularization of Science in Great Britain." *Brit. J. Hist. Sci.* 2 (1965-66): 221-234.

Argues that as dictionaries were intended to entertain as well as instruct, they are a useful indicator of the popularization of science. The example of Newtonian science in the 18th century is briefly examined here.

490. Lipski, A., "The Foundation of the Russian Academy." *Isis* 44 (1953): 349-354.

A very brief paper which describes the interest of
Peter the Great in the establishment of an Academy in
1712, the people he involved in the project (including
Christian Wolff), and the financial inducements that
were offered in order to attract foreign scholars
prior to the official opening in 1726.

491. McClellan, James E., III, "The International Organization
      of Science and Learned Societies in the Eighteenth
      Century." Ph.D. dissertation, Princeton University,
      1975.

Examines the enormous increase in numbers of societies
post-1700, and discusses how, despite difficulties,
they became the chief means of disseminating scientific
information amongst the international scientific community.

491a. McClellan, James E., III, "The Scientific Press in Transi-
      tion: Rozier's Journal and the Scientific Societies
      in the 1770s." *Ann. Sci.* 36 (1979): 425-449.

Examines the early years of Rozier's *Observations sur
la physique*, focusing on its audience, the extent of
its specialization, and Rozier's connections with the
scientific societies of the day. Points out that while
the journal showed some signs of specialization, it
nevertheless embraced the whole range of experimental
and observational sciences (including experimental
physics) and excluded only the mathematical ones. Thus
sees the journal as filling a transitional role between
the 18th-century learned society press and the specialized
periodicals of the 19th century.

492. McClellan, James E., III, "The Académie Royale des
      Sciences, 1699-1793: A Statistical Portrait." *Isis*
      72 (1981): 541-567.

A statistical survey based on the 1979 edition of the
Academy's *Index biographique* (item 9), serving to con-
firm the picture of the membership given by Hahn (item
485).

493. McKie, Douglas, "The Scientific Periodical from 1665 to
      1788." *Natural Philosophy Through the 18th Century
      and Allied Topics: Philosophical Magazine Commemora-
      tive Volume*, ed. Allan Ferguson. London, 1948. Reprint-
      ed, London: Taylor and Francis, 1972, pp. 122-132.

Describes the rapid increase in the number of journals at the end of the century, providing an account of the important *Observations sur la physique* of Rozier as the exemplar of a' new "serious" periodical.

494. McKie, Douglas, "Scientific Societies to the End of the Eighteenth Century." *Natural Philosophy through the 18th Century and Allied Topics: Philosophical Magazine Commemorative Volume*, ed. Allan Ferguson. London, 1948. Reprinted, London: Taylor and Francis, 1972, pp. 133-143.

Describes the many different types of societies, providing some account of their affiliations, membership and (often limited) achievements.

495. McKie, Douglas, "The 'Observations' of the Abbé François Rozier (1734-1793): I." *Ann. Sci.* 13 (1957): 73-89.

Gives bibliographical details of the rare early numbers of Rozier's *Observations sur la physique* (later the *Journal de physique*). Includes a transcription of one of the first printed accounts of Black's classic researches on latent heat (pp. 86-89).

496. Martin, Geneviève, "Documents de l'Académie de Rouen concernant l'enseignement des sciences au XVIIIe siècle." *Rev. Hist. Sci.* 11 (1958): 207-226.

Publishes the scanty surviving MSS from a variety of sources concerning the establishment and work of the Rouen Academy.

497. Millburn, John R., "Benjamin Martin and the Royal Society." *Notes Rec. Roy. Soc. Lond.* 28 (1973-74): 15-23.

Prints seven letters from Martin to Hans Sloane, Martin Folkes and the Duke of Richmond telling the story of his inept attempt to become F.R.S., which would "have been of such advantage to my Proffesion" as lecturer.

498. Morrell, J.B., "The University of Edinburgh in the Late 18th Century: Its Scientific Eminence and Academic Structure." *Isis* 62 (1971): 158-171.

Describes the freedom with which students were able to attend a wide range of courses given by eminent practitioners and indicates the relative attendance at the various classes at the end of the century.

499. Nentwig, Heinrich, *Die Physik an der Universität Helm-
      stedt von 1700-1810*. Wolfenbüttel: Otto Wollermann,
      1891. 46 pp.

      Gives much useful detail as to the contents of the
      physics course at Helmstedt during the period stated.

500. Palm, L.C., "Sellius and His Newtonian Teaching of
      Physics in Halle." *Janus* 64 (1977): 15-24.

      Provides some useful information on the introduction
      of lectures in experimental physics at one of the leading
      German universities during the late 1730s.

501. Robinson, Eric, "Benjamin Donn (1729-1798), Teacher of
      Mathematics and Navigation." *Ann. Sci.* 19 (1963):
      27-36.

      Describes Donn's professional life, taking him as an
      exemplar of the non-peripatetic lecturer, tied to a
      school and lecturing in the immediate locality.
      Briefly describes his courses, his other work (lecturing
      and surveying), his publications, and details of his
      remuneration.

502. Robinson, Eric, "The Lunar Society and the Improvement
      of Scientific Instruments. I." *Ann. Sci.* 12 (1956):
      296-304. "The Lunar Society and the Improvement of
      Scientific Instruments. II." *Ann. Sci.* 13 (1957):
      1-8.

      These two articles describe the close co-operation be-
      tween scientists and industrialists in the Society allow-
      ing such devices as Wedgwood's pyrometer, micrometers,
      balances and various optical instruments to be produced
      cheaply and efficiently.

503. Rosen, Richard L., "The Academy of Sciences of the In-
      stitute of Bologna, 1690-1804." Ph.D. dissertation,
      Case Western Reserve University, 1971. 354 pp.

      Discusses the origins and work of the Academy in a
      straightforward narrative history. Notes a concentration
      of interest in the Academy in topics such as electricity,
      hydrology and luminescence.

*   Ruestow, Edward G., *Physics at 17th and 18th-Century
      Leiden*.

      Cited herein as item 449.

504. Schaff, Josef, *Geschichte der Physik an der Universität
      Ingolstadt.* Erlangen: K.B. Hof- und Universitätsbuch-
      druckerei von Junge und Sohn, 1912. vi + 234 pp.

      Pp. 148-161 deal with the period 1705-1748, when
      "physics" was still understood in the old (Aristotelian)
      sense but was now taught from an atomistic point of
      view. Pp. 162-193 cover the period 1748-1800, when the
      new-style experimental physics was in the ascendancy.

505. Schofield, Robert E., *The Lunar Society of Birmingham:
      A Social History of Provincial Society and Industry
      in 18th Century England.* Oxford: Clarendon Press,
      1963. xii + 491 pp.

      The definitive history of the shadowy provincial
      society which, between 1765 and 1790, included such
      notables as Boulton, Erasmus Darwin, Keir, Priestley,
      Wedgwood and Watt. Schofield, drawing extensively on
      manuscript and archival materials, examines the researches
      of the individuals, their mutual cooperation, and their
      connection with the wider social and economic aspects of
      their time. Concludes with a sketch of the influence
      of the Society.

506. Schofield, Robert E., "The Lunar Society of Birmingham:
      A Bicentenary Appraisal." *Notes Rec. Roy. Soc. Lond.*
      21 (1966): 144-161.

      Drawing from his monograph (item 505), Schofield pro-
      vides a brief account of the rise of the Lunar Society,
      its membership, its 'formalization' about 1775, the co-
      operative work produced by its members and its decline
      following its fragmentation about 1790.

*   Taton, René, *Enseignement et diffusion des sciences en
      France au XVIIIe siècle.*

      Cited herein as item 463.

507. Taylor, F.S., "Science Teaching at the end of the 18th
      Century." *Natural Philosophy through the 18th Century
      and Allied Topics: Philosophical Magazine Commemorative
      Volume*, ed. Allan Ferguson. London, 1948. Reprinted,
      London: Taylor and Francis, 1972, pp. 144-164.

      Provides details of the education of some seventy
      prominent scientists and examples of the gradually in-
      creasing access during the century to some form of
      scientific education.

508. Torlais, Jean, "L'Académie de La Rochelle et la diffusion
      des sciences au XVIIIe siècle." *Rev. Hist. Sci.* 12
      (1959): 111-125.

      Gives an account of the origins, the membership
      (including Réaumur), the publications and the eventual
      disbanding of the La Rochelle Academy. Includes a sec-
      tion on the work of the Academy on physical subjects,
      especially its investigation of the electric fish, the
      torpedo.

                        (c) Instrumentation

                    See also section V(g)(ii)
                for works dealing with thermometry.

509. Bos, H.J.M., *Mechanical Instruments in the Utrecht Uni-
      versity Museum*. Utrecht: Utrecht University Museum,
      1968. xii + 70 pp.

      Descriptive catalogue of a large number of instruments,
      most dating from the 18th century, with a description
      of the teaching of mechanics at the university. Forty
      photographs.

*   Chenakal, V.L., "Elektricheskie mashiny v Rossii XVIII
      veka" ("Electrical machines in 18th-century Russia").

      Cited herein as item 640a.

510. Clay, Reginald Stanley, and Thomas H. Court, *The History
      of the Microscope Compiled from Original Instruments
      and Documents, Up to the Introduction of the Achromatic
      Microscope*. London: Charles Green, 1932. xiv + 266 pp.

      The development of various different types of instru-
      ment (e.g., simple, compound, Culpepper) is described
      and illustrated. Most space is given to 18th-century
      instruments. A useful but by no means complete list of
      instrument makers is included. Now somewhat dated.

511. Crommelin, C.A., *Descriptive Catalogue of the Physical
      Instruments of the 18th Century (including the Col-
      lection 'sGravesande-Musschenbroek) in the Rijksmuseum
      voor de Geschiedenis der Natuurwetenschappen at Leyden*.
      Leiden: Rijksmuseum voor de Geschiedenis der Natuur-
      wetenschappen, 1951. 74 pp.

      A very important early 18th-century collection.

512. Daumas, Maurice, *Scientific Instruments of the Seventeenth and Eighteenth Centuries and their Makers*. Translated and edited by Mary Holbrook. New York and Washington: Praeger/London: Batsford, 1972. vi + 361 pp.

A translation of Daumas' 1953 classic, *Les instruments scientifiques aux XVIIe et XVIIIe siècles*, but incorporating many corrections and with updated bibliographical material. Provides both a history of instrument development and utilization, and accounts of the makers and their workshops. Not unexpectedly, there is an emphasis on French sources and French makers. Reviewed by D. de Solla Price, *Isis* 64 (1973): 421-422 and D.J. Bryden, *Brit. J. Hist. Sci.* 7 (1974): 87-88.

* Finn, Bernard S., "Output of Eighteenth-Century Electrical Machines."

Cited herein as item 653.

513. Forbes, Eric G., "The Origin and Development of the Marine Chronometer." *Ann. Sci.* 22 (1966): 1-25.

Describes the establishing of the Parliamentary prize for a successful determination of longitude (1714), Harrison's extensive testing and development of his chronometer, the eventual award of the prize to Harrison, and the work of his successors--Mudge, Arnold and Earnshaw--to the end of the 18th century.

514. Goodison, Nicholas, "Charles Orme's Place in the History of the Barometer." *Physis* 8 (1966): 373-382.

Describes Orme's (1688-1747) innovations in barometer design, many of which later became standard.

515. Goodison, Nicholas, "Daniel Quare and the Portable Barometer." *Ann. Sci.* 23 (1967): 287-293.

Argues that Quare did not, as has been supposed, invent the first truly portable barometer.

* Hackmann, Willem D., "The Design of the Triboelectric Generators of Martinus van Marum, F.R.S."

Cited herein as item 665.

* Hackmann, Willem D., *John and Jonathan Cuthbertson: The Invention and Development of the Eighteenth-Century Plate Electrical Machine.*

Cited herein as item 666.

*   Hackmann, Willem D., *Electricity from Glass: The History
        of the Frictional Electrical Machine, 1600-1850.*

        Cited herein as item 667. Chiefly concerned with 18th-
        century developments.

*   Hackmann, Willem D., "The Relationship between Concept
        and Instrument Design in Eighteenth-Century Experimental
        Science."

        Cited herein as item 668.

516.   Hellman, C.D., "John Bird (1709-1776), Mathematical
        Instrument Maker in the Strand." *Isis* 17 (1932):
        127-153.

        Briefly describes the life of this internationally
        known craftsman and gives an account of his work, mainly
        on astronomical and navigational instruments. Provides
        details of the means whereby he divided the scales of
        his instruments accurately.

*   Henderson, Ebenezer, *Life of James Ferguson, F.R.S.*

        Cited herein as item 60.

517.   Kniajetskaia, E.A., and V.L. Chenakal, "Pierre le Grand
        et les fabricants français d'instruments scientifiques."
        Rev. Hist. Sci. 28 (1975): 243-258.

        Describes Peter the Great's involvement with science
        in his travels in Western Europe, especially his pur-
        chase of instruments, and prints three bills of sale of
        French instruments that he bought.

518.   Meadows, A.J., "Observational Defects in Eighteenth-
        Century British Telescopes." *Ann. Sci.* 26 (1970):
        305-317.

        A general account, drawn largely from contemporary com-
        ments. The discussion of defects in glass, types of
        glass and difficulties in grinding lenses apply, of
        course, to all optical apparatus.

*   Millburn, John R., *Benjamin Martin, Instrument Maker and
        "Country Showman."*

        Cited herein as item 80.

*   Prinz, H., "Erschütterndes und Faszinierendes über
        gespeicherte Elektrizität."

        Cited herein as item 695.

* Robinson, Eric, "The Lunar Society and the Improvement of Scientific Instruments."

    Cited herein as item 502.

518a. Schmitz, Rudolf, "Die physikalische Gerätesammlung der Universität Marburg im 17. und 18. Jahrhundert." *Sudhoffs Archiv* 60 (1976): 375-403.

    A chronological survey of the development of the physics "cabinet" at Marburg from its beginnings in the 1690s until the early years of the 19th century.

519. Skempton, A.W., and Joyce Brown, "John and Edward Troughton, Mathematical Instrument Makers." *Notes Rec. Roy. Soc. Lond.* 27 (1972-73): 233-262.

    Corrects the "traditional" erroneous biography of the Troughtons who between them were the "second" instrument makers in London c. 1780-1800 and the foremost from 1800 to 1825. Includes descriptions of many of their instruments, and an appendix giving details of some twenty-five surviving examples.

* Sobol', S.L., *Istoriya mikroskopa i mikroskopicheskikh issledovanii v Rossii v XVIII veke.*

    Cited herein as item 594.

520. Turner, G.L'E., "The Auction Sale of the Earl of Bute's Instruments, 1793." *Ann. Sci.* 23 (1967): 213-243.

    A transcription of the catalogue of the sale of the Earl of Bute's extensive collection of 18th-century "optical, mathematical and philosophical" instruments, giving prices and identifying the buyers listed.

521. Turner, G.L'E., *Antique Scientific Instruments.* Poole, Dorset: Blandford Press, 1980. 168 pp.

    A "popular" book, divided into sections according to different types of instruments. Each section provides a brief history of the design and utilization of the types with a listing of a few of the more common makers. An inadequate index. 48 pages of colour plates.

522. Turner, G.L'E., and T.H. Levere, *Van Marum's Scientific Instruments in Teyler's Museum. (Martinus Van Marum: Life and Work,* ed. R.J. Forbes *et al.*, vol. IV.) Leiden: Hollandsche Maatschappij der Wetenschappen and Noordhoff International Publishing, 1973. ix + 401 pp.

Deals with the collection of more than 500 instruments
set up by Van Marum between 1784 and 1837 using the be-
quest of the Dutch merchant Pieter Teyler. The first of
two parts consists of three long essays by the editors.
Ch. 1 by Turner discusses the relationship between
progress in instrument making and that of science, the
place of scientific societies in disseminating knowledge
and the importance of craftsmen and entrepreneurs in the
history of science at that time. Chs. 2 and 3 are by
Levere and deal respectively with the story of Van Marum
and the Teyler Museum, and the interaction of ideas and
instruments on Van Marum's scientific work. Part II
deals in superb detail with the instrument collection,
giving full information about the instruments and about
applications of these and similar types by various other
contemporary practitioners. Reviewed by W.F. Ryan, *Ann.
Sci.* 32 (1975): 299-300. (cf. item 50).

*   Walker, W. Cameron, "The Detection and Estimation of
        Electric Charges in the Eighteenth Century."

        Cited herein as item 706.

523.   Wheatland, David P., *The Apparatus of Science at Harvard,
        1765-1800*. Cambridge, Mass.: Harvard University Press/
        London: Oxford University Press, 1968. xi + 204 pp.

        Describes and illustrates the approximately 85 instru-
        ments, mainly by the "better" London makers, in the
        Harvard collection. Gives details of purchase and use,
        providing quotations from letters and from account books
        in the university archives, thus supplying valuable
        details of the history of the instruments and the part
        they played in the development of science at Harvard.

        (d) Foundational Issues, Properties of Matter, Sound

524.   Barkla, H.M., "Benjamin Robins and the Resistance of
        Air." *Ann. Sci.* 30 (1973): 107-122.

        Describes the means (including Robins' own invention,
        the ballistic pendulum) by which he measured air re-
        sistance between 1742 and 1751, compares his results
        with modern values, and identifies sources of error.

525. Bikerman, J.J., "Capillarity before Laplace: Clairaut, Segner, Monge, Young." *Arch. Hist. Exact Sci.* 18 (1977-78): 103-122.

Describes the work of the four authors mentioned, finding little of significance in their contributions compared to those of Laplace.

526. Birembaut, Arthur, "Les deux déterminations de l'unite de masse du système métrique." *Rev. Hist. Sci.* 12 (1959): 25-54.

A brief history of pre-metric systems of weight in France and an account of the definition of the *grave* by Lavoisier and Haüy, and its replacement by the *kilogramme*.

527. Briggs, J. Morton, Jr., "Aurora and Enlightenment: Eighteenth-Century Explanations of the Aurora." *Isis* 58 (1967): 491-503.

Describes the theories of Halley, Mairan and Euler, which relied respectively on magnetic particles and the solar and terrestrial atmospheres, and also others utilizing exhalations and electricity. Uses these to illustrate the limitations of the prevailing "mechanical" paradigm.

528. Burke, John G., *Origins of the Science of Crystals.* Berkeley/Los Angeles: University of California Press, 1966. 198 pp.

Devoted largely to late 18th and early 19th-century crystallography, especially Haüy's work, and concerned mostly with the development of ideas on structure. Has a succinct treatment of double refraction and other optical properties, but unfortunately little on pyro-electricity. Reviewed by R. Hooykaas, *Isis* 58 (1967): 568-570.

528a. Busch, Hermann Richard, *Leonhard Eulers Beitrag zur Musiktheorie.* Regensburg: G. Bosse, 1970. 141 pp.

A systematic analysis of Euler's *Tentamen novae theoriae musicae* (1739) and of a number of lesser later writings on music.

528b. Cohen, Albert, *Music in the French Royal Academy of Sciences: A Study in the Evolution of Musical Thought.* Princeton, N.J.: Princeton University Press, 1981. xviii + 150 pp.

Surveys the Academy's involvement in the study of music
and the theory of sound in the period from its foundation
to the Revolution. At first, the Academy's interest was
of the traditional kind, namely in music as a branch of
mixed mathematics. Later, it also became concerned in
questions of musical practice, especially in relation to
innovations in instrument design. Later still, its interest waned as music came to be seen no longer as an
integral part of physical science. Pays particular
attention to the contributions of Joseph Sauveur (1653-
1716).

529. Dyment, S.A., "Some Eighteenth Century Ideas concerning
      Aqueous Vapour and Evaporation." *Ann. Sci.* 2 (1937):
      465-473.

      Shows that in the first half of the 18th century most
      theories of evaporation conjectured either that water
      particles became joined with other, lighter particles
      (whether of air, fire, "weightless" matter, or elec-
      tricity) or that heat caused expansion of the droplets
      until they were less dense than air. In mid-century
      came the theory that the water dissolved in the air.
      Finally, water vapour came to be seen as an elastic
      fluid in its own right.

530. Hooykaas, R., "The First Kinetic Theory of Gases."
      Arch. Int. Hist. Sci. 2 (1948): 180-184.

      A brief description of Euler's 1727 paper deriving
      the value of the elasticity of the air from a complicated
      vortex model of air particles.

530a. Hooykaas, R., "La cristallographie dans l'*Encyclopédie*."
      Rev. Hist. Sci. 4 (1951): 344-352. Reprinted, *L'"Ency-
      clopédie" et le progrès des sciences et des techniques*,
      ed. S. Delorme and R. Taton. Paris: Presses Universi-
      taires de France, 1952, pp. 141-149.

      A brief review of the various articles in the *Encyclo-
      pédie* dealing with crystals. Emphasizes how far removed
      mid-18th-century crystallography was from the exact
      science developed later in the century by Haüy.

530b. Hooykaas, R., "Torbern Bergman's Crystal Theory." *Lychnos*
      (1952): 21-54.

      Analyzes Bergman's ideas in detail before considering
      the extent of Haüy's indebtedness to him. Concludes that
      "Haüy's first publications not only refer to Bergman's

work, but also show an essential affinity with it, however much they surpass it by correlating various forms
in a more correct manner."

530c. Metzger, Hélène, *La genèse de la science des cristaux.*
Paris, 1918. Reprinted, Paris: Librairie Albert
Blanchard, 1969. 248 pp.

A detailed study of the emergence of crystallography
as an independent science in the period between Steno's
work in the 1660s and Haüy's at the end of the 18th century.

531. Millington, E.C., "Studies in Capillarity and Cohesion in
the Eighteenth Century." *Ann. Sci.* 5 (1941-47): 352-
369.

Describes the extensive experimental work done on these
subjects by Carré, Hauksbee, Taylor, Jurin, Desaguliers,
Musschenbroek and Monge, and the influential theories
of matter and cohesion of Newton and Boscovich.

* Suchting, W.A., "Euler's 'Reflections on Space and Time.'"

Cited herein as item 563.

* Truesdell, Clifford A., "The Theory of Aerial Sound,
1687-1788."

Cited herein as item 567.

(e) Mechanics, Fluid Mechanics

532. Aiton, E.J., *The Vortex Theory of Planetary Notions.*
London: Macdonald/New York: American Elsevier, 1972.
x + 282 pp.

Contains an account of Cartesian vortex theory in
France 1700-1729, the introduction of Newton's ideas
in the same period, the brief attempt to reconcile
these two theories, and the final defeat of the vortex
theories. Discusses in particular the ideas of the
Bernoullis, Malebranche, Varignon, 'sGravesande,
Bouguer and Maupertuis.

* Bos, H.J.M., *Mechanical Instruments in the Utrecht University Museum.*

Cited herein as item 509.

533. Bos, H.J.M., "Mathematics and Rational Mechanics." *The*
      Ferment of Knowledge: Studies in the Historiography
      of Eighteenth-Century Science, ed. G.S. Rousseau and
      Roy Porter. Cambridge: Cambridge University Press,
      1980, pp. 327-355.

      An excellent survey of recent work on the mathematical
      sciences in the 18th century. Truesdell's work on the
      rational mechanics of the age is deservedly given great
      prominence, while the idiosyncrasies of his approach are
      at the same time treated very judiciously.

534. Brunet, Pierre, *Etude historique sur le principe de la*
      moindre action. Paris: Hermann, 1938. 114 pp.
      (Actualités scientifiques et industrielles, 693).

      Covers the history of the principle of least action
      from its origin in Maupertuis' work, his application of
      it to mechanics and optics, criticisms of his position,
      and finally the use of the principle by Euler, Lagrange,
      Carnot and Poisson.

*   Calinger, Ronald S., "The Newtonian-Wolffian Confronta-
      tion in the St. Petersburg Academy of Sciences (1725-
      1746)."

      Cited herein as item 362a.

535. Cannon, John T., and Sigalia Dostrovsky, *The Evolution*
      of Dynamics: Vibration Theory from 1687 to 1742. New
      York/Heidelberg/Berlin: Springer-Verlag, 1981. x +
      184 pp.

      A technical account of the development during the
      period stated of the science of dynamics as applicable
      to mechanical systems with several degrees of freedom.
      Certain passages seen as central to the development of
      the subject, written by various authors from Newton to
      Johann Bernoulli, are summarized using modernized nota-
      tion and embellished with linking contextual and his-
      torical remarks. An appendix reprints in facsimile
      Daniel Bernoulli's papers from the 1730s on the hanging
      chain and the linked pendulum.

536. Cardwell, D.S.L., "Some Factors in the Early Development
      of the Concepts of Power, Work and Energy." *Brit. J.*
      Hist. Sci. 3 (1966-67): 209-224.

      Argues that the concept of work, and thus those of
      energy and power, evolved largely from theoretical and

empirical aspects of engineering in Britain and France
through the 18th and early 19th centuries.

537. Chandler, Philip, "Clairaut's Critique of Newtonian At-
      traction: Some Insights into his Philosophy of Science."
      Ann. Sci. 32 (1975): 369-378.

      An examination of the controversy between Buffon and
      Clairaut over Clairaut's proposed revision of the inverse
      square law casts light on the latter's brand of New-
      tonianism--"a delicate balance between the claims of
      mathematics and those of observation."

538. Cohen, I. Bernard, "The French Translations of Isaac
      Newton's *Philosophiae Naturalis Principia Mathematica*
      (1756, 1759, 1966)." *Arch. Int. Hist. Sci.* 21 (1968):
      260-290.

      Full bibliographical details of the surviving copies
      of the 1756 edition of Mme. du Châtelet's translation
      (revised by Clairaut), the "major edition" of 1759, and
      the "pseudo-facsimile" reprint of 1966.

539. Costabel, Pierre, "Le *De viribus vivis* de R. Boscovic
      ou de la vertu des querelles de mots." *Arch. Int.
      Hist. Sci.* 14 (1961): 3-12.

      An analysis of Boscovich's first work on mechanics
      (1745) in which he applied his force law to the *vis viva*
      debate.

540. Costabel, Pierre, "Histoire du moment d'inertie." *Rev.
      Hist. Sci.* 3 (1950): 315-336.

      Concentrates mainly on the work of Euler and Lagrange.

541. Costabel, Pierre, "La méchanique dans l'*Encyclopédie*."
      Rev. Hist. Sci. 4 (1951): 267-293. Reprinted,
      *L'"Encyclopédie" et le progrès des sciences et des
      techniques*, ed. S. Delorme and R. Taton. Paris:
      Presses Universitaires de France, 1952, pp. 64-90.

      Describes both the implicit and the explicit content
      of the articles concerning mechanics in the *Encyclopédie*.

542. Costabel, Pierre, "'sGravesande et les forces vives ou
      des vicissitudes d'une expérience soi-disant cruciale."
      Mélanges Alexandre Koyré. Paris: Hermann, 1964,
      vol. 1, pp. 117-134.

Examines 'sGravesande's contributions to the *vis viva* debate in the various editions of his *Eléments de physique* from the point of view of both theory and his search for a critical experiment to decide the controversy.

543. Delsedime, Piero, "La disputa delle corde vibranti ed una lettera inedita di Lagrange a Daniel Bernoulli." *Physis* 13 (1971): 117-146.

Incorporates two unpublished letters (1759) from Lagrange to Bernoulli discussing the controversy which also included d'Alembert and Euler.

544. Fleckenstein, Joachim Otto, "Pierre Varignon und die mathematischen Wissenschaften im Zeitalter des Cartesianismus." *Arch. Int. Hist. Sci.* 2 (1948-49): 76-138.

Includes (pp. 82-110) a detailed discussion of Varignon's writings on statics and dynamics.

545. Fraser, Craig Graham, "The Approach of Jean d'Alembert and Lazare Carnot to the Theory of a Constrained Dynamical System." Ph.D. thesis, University of Toronto, 1981.

Not seen.

546. Gaukroger, Stephen, "The Metaphysics of Impenetrability: Euler's Conception of Force." *Brit. J. Hist. Sci.* 15 (1982): 132-154.

Examines Euler's attempts to restructure the science of mechanics upon necessary and self-evident principles. Focuses especially on Euler's conception of "force" and his proposed reduction of this to the impenetrability and inertia of matter.

547. Gliozzi, Mario, "Teoremi meccanici di Vincenzo Riccati." *Physis* 9 (1967): 293-300.

Describes how Riccati gave a new proof of the parallelogram of forces rule, and showed, in 1746, how the work of the resultant of two forces equals the sum of the work of the two forces.

548. Grattan-Guinness, Ivor, *The Development of the Foundations of Mathematical Analysis from Euler to Riemann.* Cambridge, Mass., and London: M.I.T. Press, 1970. xii + 186 pp.

Includes (chapter 1, pp. 1-21) a discussion of the mathematical problems engendered by the 18th-century debates on the equation of motion of a vibrating string. Discusses the work of d'Alembert, Euler, Daniel Bernoulli and Lagrange.

*   Guenther, Siegmund, "Note sur Jean-André de Segner, fondateur de la métérologie mathématique."

Cited herein as item 387.

549. Hahn, Roger, *L'hydrodynamique au XVIIIe siècle: aspects scientifiques et sociologiques*. Paris: Palais de la Découverte, 1965. 27 pp.

A brief discussion of the development of hydrodynamics following the advances in both the theoretical mechanics and practical engineering aspects of the discipline, and an examination of some parts of their interrelationship.

550. Hall, A. Rupert, "Cartesian Dynamics." *Arch. Hist. Exact Sci.* 1 (1960-62): 172-178.

Ridicules Malebranche's attempt to explain both planetary motion and surface gravitational effects with inadequate mathematics and a mechanism of interlocking layered vortices.

551. Hankins, Thomas L., "Eighteenth-Century Attempts to Resolve the *vis-viva* Controversy." *Isis* 56 (1965): 281-297.

By examining the reasons given by 'sGravesande, d'Alembert and Boscovich for dismissing the controversy as a "mere matter of words," Hankins shows that much of the problem was due to confusions over the meaning of "force" (associating it with a variety of other mechanical quantities such as work and pressure) and difficulties due to people holding different theories of matter.

552. Hankins, Thomas L., "The Reception of Newton's Second Law of Motion in the Eighteenth Century." *Arch. Int. Hist. Sci.* 20 (1967): 43-65.

Shows the limitations and ambiguities in Newton's formulations of his second law, and the philosophical and physical problems it presented to theoreticians such as Euler, d'Alembert, Maupertuis, Lazare Carnot and Varignon. Describes alternative formulations of the law, culminating in 1750 with Euler's $F = m \dfrac{d^2x}{dt^2}$.

553. Hankins, Thomas L., "The Influence of Malebranche on the Science of Mechanics during the Eighteenth Century." *J. Hist. Ideas* 28 (1967): 193-210.

Shows that criticisms of his Cartesianism and theology notwithstanding, Malebranche moderated the influence of Newtonianism on authors such as Euler, Maupertuis, Lazare Carnot and especially d'Alembert, first in advocating a rational rather than an experimental mechanics, secondly in his "law of the simplest means," and finally in his criticism of the concept of force, which led either to its being used purely pragmatically or to its being rejected outright from mechanics.

*   Hankins, Thomas L., *Jean d'Alembert: Science and the Enlightenment*.

Cited herein as item 396.

553a. Heimann, P.M., and J.E. McGuire, "Cavendish and the *vis viva* Controversy: A Leibnizian Postscript." *Isis* 62 (1971): 225-227.

The authors publish a brief, previously unpublished manuscript of Cavendish's, dating from about 1760, on the definition of "force." Like d'Alembert before him, Cavendish concludes that the *vis viva* controversy was merely a dispute about words. The authors stress the importance of Cavendish's accepting the Leibnizian conception of *vis viva* as a conserved quantity.

554. Heimann, P.M., "'Geometry and Nature': Leibniz and Johann Bernoulli's Theory of Motion." *Centaurus* 21 (1977): 1-26.

Argues that while Bernoulli's theory of motion (1727) was based on Leibnizian dynamics, it differed from Leibnizian natural philosophy in that, via an analogy between mathematics and nature and the "law of continuity," he assigned a real physical status to infinitesimal *vis mortua*, and treated *vis viva* as a physical entity, which Leibniz did not.

555. Hiebert, Erwin N., *Historical Roots of the Principle of Conservation of Energy*. Madison, 1964. Reprinted, New York: Arno Press, 1981. 118 pp.

Traces the antecedents of the 19th-century conservation principle, as a principle of the science of mechanics, from its earliest beginnings through the 18th century.

556. Iltis, Carolyn, "D'Alembert and the *vis viva* Controversy." *Studs. Hist. Phil. Sci.* 1 (1970): 135-144.

Shows that d'Alembert contributed nothing to the *vis viva* debate that had not been anticipated by 'sGravesande (that it was a dispute over words) or Boscovich (that *vis viva* was a measure of force over distance, and momentum of force over time), and that even these recapitulations did not occur in the first (1743) edition of the *Traité de dynamique* but only in the second (1758) edition.

557. Iltis, Carolyn, "The Decline of Cartesianism in Mechanics: The Leibnizian-Cartesian Debates." *Isis* 64 (1973): 356-373.

Describes the papers of Pierre Mazière, Jacques de Louville, Jean Bernoulli, Jean Jacques de Mairan, Jean-Pierre de Crousaz and Charles-Etienne Camus that were entered for competitions set by the Paris Academy of Sciences in the 1720s. Shows how, starting from Cartesian metaphysical or mechanical presuppositions, these men were unable to arrive at a consistent Cartesian mechanics, but were able at best to achieve results that supported Newtonian or Leibnizian mechanics.

558. Iltis, Carolyn, "Madame du Châtelet's Metaphysics and Mechanics." *Studs. His. Phil. Sci.* 8 (1977): 29-48.

Classifies du Châtelet's *Institutions de physique* (1740) as Newtonian in basic mechanics, Leibnizian in dynamics, and an integration of Leibniz, Descartes and Newton in natural philosophy. Sees her as representative of a general integrating trend in the 1740s, but also shows that she made little contribution to clarifying the *vis viva* controversy, being stolidly Leibnizian on this question.

559. Laudan, Laurens L., "The *vis viva* Controversy: A Post-Mortem." *Isis* 59 (1968): 131-143.

Argues convincingly that, contrary to legend, d'Alembert's *Traité de dynamique* of 1743 neither ended the *vis viva* controversy nor even had any significant influence on those who reached similar conclusions. Also shows that the majority of writers *circa* 1760 believed, contrary to d'Alembert, that momentum was the true measure of force.

560.   Pacey, A.J., and S.J. Fisher, "Daniel Bernoulli and the
       vis viva of Compressed Air." *Brit. J. Hist. Sci.* 3
       (1966-67): 388-392.

       This short note draws attention to a passage in the
       Hydrodynamica (1738) where Bernoulli calculates the
       vis viva potentialis stored in compressed air. The
       authors argue that the concepts deployed in this calcu-
       lation anticipate the much later distinction between
       kinetic and potential energies.

560a. Palter, Robert, "Kant's Formulation of the Laws of
       Motion." *Synthese* 24 (1972): 96-116.

       An analysis of Kant's discussion of the laws of
       motion in his *Metaphysical Foundations of Natural
       Science*.

561.   Papineau, David, "The *vis viva* Controversy: Do Meanings
       Matter?" *Studs. Hist. Phil. Sci.* 8 (1977): 111-142.

       Argues that the controversy must be seen in terms of
       two competing systems of ideas, both of which accepted
       Descartes' concept that "force" is conserved and merely
       transferred in impact, but which used different measures
       of force. The extension and justification of their
       particular measures, with their concomitant theories
       of matter and of impact mechanisms, enlarged the dis-
       pute. Finally argues that the dispute was only concluded
       when "force of motion" acquired a new role with Bosco-
       vich and d'Alembert.

*   Robinet, André, *Malebranche de l'Académie des Sciences:
       l'oeuvre scientifique, 1674-1715.*

       Cited herein as item 443a.

562.   Speziali, Pierre, "Une correspondance inédite entre
       Clairaut et Cramer." *Rev. Hist. Sci.* 8 (1955):
       193-237.

       While the correspondence (from 1729 to 1750) is large-
       ly concerned with mathematics and lunar motion, it con-
       tains many interesting asides on contemporaries of the
       two and shows the resistance that Newton's lunar theory
       still encountered.

563.   Suchting, W.A., "Euler's 'Reflections on Space and Time.'"
       Scientia 104 (1969): 270-278.

       A summary of Euler's defense of a realist view of space

and time against the relationalist doctrines of "meta-physicians" such as Leibniz, Wolff and Berkeley.

564. Tonelli, Giorgio, "La nécessité des lois de la nature au XVIIIe siècle et chez Kant en 1762." *Rev. Hist. Sci.* 12 (1959): 225-241.

Concerned primarily with laws of motion, looking at the writings of Jean Bernoulli, d'Alembert and Mauper-tuis among others, as well as Kant's early writings.

565. Truesdell, Clifford A., "Rational Fluid Mechanics, 1687-1765." Editor's Introduction to *Leonhardi Euleri Opera Omnia*, Ser. 2, vol. 12. Zürich: Orell Füssli, 1954, pp. ix-cxxv.

Follows a chronological sequence highlighting Euler's contributions to the field. Examines prior work that Euler drew on, and analyzes his early work and that of his contemporaries, the Bernoullis. Thus meticulously traces the slow process by which Euler, drawing on the work of his predecessors, forged his mathematical physics with ever increasing thoroughness over thirty years. Uses modern mathematical notation and does not discuss physical aspects of the theories.

566. Truesdell, Clifford A., "The First Three Sections of Euler's Treatise on Fluid Mechanics, 1766." Editor's Introduction, Part 1, in *Leonhardi Euleri Opera Omnia*, Ser. 2, vol. 13. Zürich: Orell Füssli, 1955, pp. x-xviii.

Analyzes Euler's great treatise, explaining his arguments in the light of past work by both himself and others, and as an anticipation of future work in this area.

567. Truesdell, Clifford A., "The Theory of Aerial Sound, 1687-1788." Editor's Introduction, Part 2, in *Leonhardi Euleri Opera Omnia*, Ser. 2, vol. 13. Zürich: Orell Füssli, 1955, pp. xix-lxxii.

Describes the unravelling of the basic theory of acoustic propagation in a perfect medium, tracing this from Newton's first formulation of the problem to the end of the 18th century. Emphasis is placed on the analysis of Euler's neglected mathematical theories, arguing that essentially all the century's achievements were contained in his final wave equations. Explicitly avoids any concern with physical, non-mathematical, theories.

568. Truesdell, Clifford A., "Rational Fluid Mechanics, 1765-
      88." Editor's Introduction, Part 3, in *Leonhardi
      Euleri Opera Omnia*, Ser. 2, vol. 13. Zürich: Orell
      Füssli, 1955, pp. lxxiii-cv.

      Continues in the same detail Truesdell's earlier dis-
      cussions, covering the final scattered work of Euler
      and that of Borda, d'Alembert and Lagrange, the last two
      of which are chiefly noted by Truesdell for their errors
      and their down-grading of the work of the Basle school.
      Again restricts attention to the mathematics, omitting
      consideration of the physics involved.

569. Truesdell, Clifford A., *The Rational Mechanics of Flex-
      ible or Elastic Bodies, 1638-1788*. Editor's Introduc-
      tion to *Leonhardi Euleri Opera Omnia*, Ser. 2, vols.
      10 and 11, in *Leonhardi Euleri Opera Omnia*, Ser. 2,
      vol. 11, part 2. Zürich: Orell Füssli, 1960. 435 pp.

      A monumental work, divided into pre-Eulerian, Eulerian
      and post-Eulerian work (the latter including especially
      d'Alembert, Coulomb and Chladni), the main emphasis being
      on the Basle school comprising Euler and the Bernoullis.
      Traces the history of the twin problems of the static
      and dynamical behavior of deformable bodies, presenting
      this in terms of more or less successful efforts to solve
      particular problems of increasing generality. Also
      mentions experimental work in the area. Excellent sum-
      maries at the end of sections. The mathematical notation
      is modernized. Reviewed by J.R. Ravetz, *Isis* 53 (1962):
      263-266.

570. Truesdell, Clifford A., "A Program toward Rediscovering
      the Rational Mechanics of the Age of Reason." *Arch.
      Hist. Exact Sci.* 1 (1960-62): 1-36. Reprinted, *Essays
      in the History of Mechanics* by C.A. Truesdell. Berlin,
      Heidelberg and New York: Springer-Verlag, 1968, pp.
      85-137.

      A brief description of the more important developments
      in the science of rational mechanics from its foundations
      in the specific problems of Newton's *Principia* (1687)
      and in Jean Bernoulli's work, through the development
      of techniques and solutions of increasing generality
      concerning the equations of motion, equilibrium, elas-
      ticity and stress, to Lagrange's *Mécanique analytique*
      (1788). Concentrates on the work of d'Alembert, the
      Bernoullis, and Euler.

571. Truesdell, Clifford A., *Essays in the History of Mechanics*. Berlin, Heidelberg and New York: Springer-Verlag, 1968. xii + 384 pp.

Relevant essays are described separately in this bibliography (items 570, 572, 573, 574).

572. Truesdell, Clifford A., "Reactions of Late Baroque Mechanics to Success, Conjecture, Error, and Failure in Newton's *Principia*." *Essays in the History of Mechanics* by C.A. Truesdell. Berlin, Heidelberg and New York: Springer-Verlag, 1968, pp. 138-183. Also published in *The annus mirabilis of Sir Isaac Newton, 1666-1966*, ed. R. Palter. Cambridge, Mass.: M.I.T. Press, 1971, pp. 192-232.

A general account of 18th-century rational mechanics, presenting this as a response to both the originality and the inadequacies of Book II of Newton's *Principia*. Argues that the successes of the programme should be seen in terms of the discovery of solutions to particular problems, mainly by d'Alembert and the Basle School (i.e., Euler and the Bernoullis), who drew largely on principles that Newton had ignored and who worked outside any "Newtonian" tradition.

573. Truesdell, Clifford A., "The Creation and Unfolding of the Concept of Stress." *Essays in the History of Mechanics* by C.A. Truesdell. Berlin, Heidelberg and New York: Springer-Verlag, 1968, pp. 184-238.

Traces the history of the stress principle--the foundation of continuum mechanics--prior to its first enunciation by Cauchy. Argues that its origins lie in the solution of diverse problems of mechanics concerning tension, hydraulics and hydrodynamics, especially by the Bernoullis and Euler.

574. Truesdell, Clifford A., "Whence the Law of Moment of Momentum?" *Essays in the History of Mechanics* by C.A. Truesdell. Berlin, Heidelberg and New York: Springer-Verlag, 1968, pp. 239-271. Previously published in *Mélanges Alexandre Koyré*. Paris: Hermann, 1964. Vol. I, pp. 149-158.

Traces the derivation of the law, now seen as one of the fundamental principles of mechanics, in the work of Jakob and Daniel Bernoulli and (especially) Euler. Conjectures that the source of the "physicist's" version of the law is Poisson's work of 1833.

575. Vassails, Gérard, "L'*Encyclopédie* et la physique."
      Rev. Hist. Sci. 4 (1951): 294-323. Reprinted, *L'"Ency-
      clopédie" et le progrès des sciences et des techniques*,
      ed. S. Delorme and R. Taton. Paris: Presses Universi-
      taires de France, 1952, pp. 91-120.

      Shows the emphasis on mechanics and geometrical optics
      rather than the experimental sciences in articles in
      the *Encyclopédie*, and examines the theoretical and
      metaphysical background to these articles.

                            (f) Light

576. Badcock, A.W., "Physical Optics at the Royal Society,
      1660-1800." *Brit. J. Hist. Sci.* 1 (1962): 99-116.

      Discusses modifications made to Newton's corpuscular
      theory of light in order to account for total internal
      reflection, dispersion, interference, diffraction and
      phosphorescence, and describes attempts to determine
      the velocity of light.

577. Blay, Michel, "Une clarification dans le domaine de
      l'optique physique: 'bigness' et promptitude." *Rev.
      Hist. Sci.* 33 (1980): 215-224.

      Argues that Newton's notion of vibrations of varying
      "bigness" as a possible explanation of the different
      colours of light remained rooted in physiological optics,
      whereas Malebranche's conception of "promptitude" genu-
      inely foreshadowed the idea of frequency.

578. [Bouguer, Pierre], *Pierre Bouguer's Optical Treatise on
      the Gradation of Light*. Translated with introduction
      and notes by W.E. Knowles Middleton. Toronto: Univer-
      sity of Toronto Press/London: Oxford University Press,
      1962. xv + 241 pp.

      A translation of Bouguer's scarce *Essai d'optique sur
      la gradation de la lumière* (1729), a landmark in the
      history of photometry.

579. Buchwald, Jed Z., "Experimental Investigations of Double
      Refraction from Huygens to Malus." *Arch. Hist. Exact
      Sci.* 21 (1979-80): 311-373.

      Exhaustively traces 18th-century attempts to determine
      a law of double refraction experimentally, and attributes

the several alternative laws which arose (despite the
eventual acceptance of Huygens' original construction)
to the difficulty of achieving adequate experimental
technique and accuracy, to the geometrical expression
of the theory which restricted its generality, and
flowing from these, to the apparent conflict of experimental results with theory.

580. Cantor, G.N., "Berkeley, Reid and the Mathematization
        of Mid-Eighteenth-Century Optics." *J. Hist. Ideas*
        38 (1977): 429-448.

      Examines the influence in the 18th century of both
      Berkeley's criticism of the mathematization of geometrical
      optics and his own "sensationalist" alternative. Con-
      cludes with a discussion of Thomas Reid's absorption of
      Berkeley's ideas into a "common-sense" empiricist
      position.

581. Cantor, G.N., "The Historiography of 'Georgian' Optics."
        Hist. Sci. 16 (1978): 1-21.

      Argues that histories of 18th and early 19th-century
      optics have been overly influenced by Whewell's Whiggish
      and didactic account which maintains that the period
      was one of sterility dominated by the corpuscular theory.
      Instead, Cantor argues, this was a period of great fer-
      tility covering many aspects of optical theory and
      utilizing several diverse models for light. A critique
      is provided of recent secondary literature on this sub-
      ject, especially that of Steffens (item 596).

582. Gliozzi, Mario, "La polemica sulla fosforescenza tra
        Giambatista Beccaria e Benjamin Wilson." *Physis* 3
        (1961): 113-124.

      Concerns the controversy between Beccaria and Wilson
      in 1775-76 on the relative colours of phosphorescent
      and incandescent light. Includes an informative, pre-
      viously unprinted letter from João Jacinto Magellan to
      Beccaria.

583. Gruner, Shirley M., "Goethe's Criticism of Newton's
        Opticks." *Physis* 16 (1974): 66-82.

      Uses Goethe's criticisms to "throw a significant light
      on the non-empirical basis of Newton's science of optics."

*   Guerlac, Henry, "Newton in France: The Delayed Acceptance
        of His Theory of Color."

      Cited herein in item 392.

584.   Hakfoort, C., "Nicolas Béguelin and His Search for a
       Crucial Experiment on the Nature of Light (1772)."
       Ann. Sci. 39 (1982): 297-310.

       Describes an interesting attempt by Béguelin to find
       a method of finally deciding between the Newtonian
       emission theory of light and Euler's wave-theory alterna-
       tive. Draws attention in this way to the fact that,
       at least in Germany, the Newtonian view did not hold
       undisputed sway at this period but that, on the contrary,
       there was a lively debate going on between adherents of
       the two opposing theories.

*   Hildebrandsson, H. Hildebrand, and C.W. Oseen, *Samuel
       Klingenstiernas Levnad och Werk: Biografisk Skildring*.

       Cited herein as item 63. Includes a number of documents
       relating to the invention of achromatic lenses.

585.   Jones, H.W., "Lomonosov's Hypothesis on Light and Heat."
       *Proceedings, 13th International Congress of the History
       of Science, Moscow, 1971* (Moscow, 1974), vol. 6, pp.
       316-321.

       A brief description of Lomonosov's somewhat old-
       fashioned "mechanical" theory of heat and light.

586.   Middleton, W.E. Knowles, "Bouguer, Lambert, and the
       Theory of Horizontal Visibility." *Isis* 51 (1960):
       145-149.

       Demonstrates that the theory of the horizontal visual
       range of objects in the atmosphere was first expounded
       by Bouguer in his posthumously published *Traité d'optique*
       (1760) and later restated by Lambert (probably, according
       to Middleton, after reading the *Traité*) in 1774.

586a. Middleton, W.E. Knowles, "Archimedes, Kircher, Buffon,
       and the Burning-Mirrors." *Isis* 52 (1961): 533-543.

       Discusses Buffon's successful demonstration in 1747
       of the possibility of constructing large mirrors such
       as Archimedes is reputed to have used in defending
       Syracuse from the Roman fleet, and his refutation
       thereby of Descartes' dismissal of the story about
       Archimedes on grounds of optical theory.

587.   Morère, Jean-Edouard, "La photométrie: Les sources de
       1'*Essai d'optique sur la gradation de la lumière de*
       Pierre Bouguer, 1729." *Rev. Hist. Sci.* 18 (1965):
       337-384.

Identifies previous theories, problems and techniques
that Bouguer brought together in his photometry. Re-
jects suggestions that his work was in any real way
anticipated.

588. Nash, Leonard K., *Plants and the Atmosphere.* Cambridge,
      Mass.: Harvard University Press, 1952. ix + 122 pp.
      (Harvard Case Histories in Experimental Science,
      case 5.)

      While brief, this is one of the only histories of
      photosynthesis, the clarification of which was one of
      the major achievements of the 18th-century science of
      light. It traces the discovery of the phenomenon and
      theories concerning it through the work of Hales,
      Priestley, Ingenhousz, Senebier and de Saussure, in-
      volving various conceptions of the nature of light and
      its interaction with matter in chemical and biological
      phenomena.

589. Nordenmark, N.V.E., and J. Nordström, "Om uppfinningen
      av den akromatiska och aplanatiska linsen." *Lychnos*
      3 (1938): 1-52; 4 (1939): 313-384.

      A thorough study of Klingenstierna's important role
      in the development of achromatic lenses with major
      primary sources printed. Article in Swedish, with
      English summary; sources in original languages.

590. Pav, Peter Anton, "Eighteenth Century Optics: The Age
      of Unenlightenment." Ph.D. dissertation, Indiana Uni-
      versity, 1964.

      Sees 18th-century optics as sterile. Argues that
      this was due to the dominance of Newton and the
      mechanical philosophy and the lack of an adequate al-
      ternative, and because optics became submerged in the
      wider debate between Cartesianism and Newtonianism.

591. Pav, Peter Anton, "Louis Carré (1663-1711) and Mechanistic
      Optics." *Arch. Int. Hist. Sci.* 24 (1974): 340-348.

      Shows the grip of the mechanistic philosophy on "the
      thoroughly common member of the Academy" Carré, who,
      although his experiments of 1702-1705 intended to verify
      the Cartesian ballistic model of light were clearly
      negative, remained undaunted in his adherence to Des-
      cartes' theory.

591a. Pedersen, Kurt Møller, "Roger Joseph Boscovich and John
      Robison on Terrestrial Aberration." *Centaurus* 24 (1980):
      335-345.

A brief account of Boscovich's discussion of terrestrial aberration observed through a water-filled telescope and Robison's critical analysis of Boscovich's ideas.

*   Pfeiffer, H., "Les études perspectives et chromatologiques de Lambert."

Cited herein in item 418.

592. Priestley, Joseph, *The History and Present State of Discoveries relating to Vision, Light, and Colours.* 2 vols. London: J. Johnson, 1772.

Despite its bias towards English work of the period, this is an essential source for any study of 18th-century optics as it contains a wealth of material not easily obtainable elsewhere.

*   Robinet, André, *Malebranche de l'Académie des Sciences: l'oeuvre scientifique, 1674-1715.*

Cited herein as item 443a.

593. Schagrin, Morton L., "Early Observations and Calculations on Light Pressure." *Amer. J. Phys.* 42 (1974): 927-940.

Describes a large number of 18th-century experiments and theories to measure light pressure and shows the imperviousness of the various theories to experimental refutation. Mentions Euler, Franklin, Mairan, Musschenbroek, Priestley and Michell.

594. Sobol', S.L., *Istoriya mikroskopa i mikroskopicheskikh issledovanii v Rossii v XVIII veke.* [*History of the microscope and microscopical research in Russia in the 18th century*]. Moscow/Leningrad: Izdatel'stvo Akademii Nauk S.S.S.R., 1949. 606 pp.

An encyclopedic work that provides much detailed information on Russian optical craftsmen as well as on Russian innovations in microscope design. Includes discussions of the important work of Euler, Aepinus and others on the design of achromatic microscopes.

595. Speiser, David, "The Distance of the Fixed Stars and the Riddle of the Sun's Radiation." *Mélanges Alexandre Koyré.* Paris: Hermann, 1964, vol. 1, pp. 541-551.

Describes Euler's attempts to accommodate Bouguer's work on light intensity into his wave theory of light with reference to the above-mentioned specific problems.

*   Speiser, David, "L'oeuvre de Lambert dans le domaine de
        l'optique."

        Cited herein in item 418.

596.   Steffens, Henry John, *The Development of Newtonian Op-
        tics in England*. New York: Science History Publica-
        tions, 1977. viii + 190 pp.

        Surveys the evolution of the corpuscular theory of
        light from Newton to the work of David Brewster in the
        early 19th century. So far as the intervening period
        is concerned, pays particular attention to the writings
        of Robert Smith, Melville, Robison and Brougham. Also
        includes a chapter on Young's development of the wave
        theory alternative. Reviewed by R.W. Home, *Ann. Sci.*
        35 (1978): 650-652. Cf. also item 581.

597.   Wells, George A., "Goethe's Qualitative Optics." *J.
        Hist. Ideas* 32 (1971): 617-626.

        Describes Goethe's theory that colours were due to a
        mixing of light and dark in a suitably turbid medium,
        and his criticism of mathematical theories on the
        grounds that the mathematics was irrelevant to reality.

598.   Wilde, Emil, *Geschichte der Optik*. 2 vols. Berlin:
        Rucker and Puchler, 1838-43.

        Only the first two volumes of a planned 3-volume work
        were ever published. Vol. 1 covers the period from
        Aristotle to the mid-17th century, vol. 2 continues the
        story from Newton, Huygens and Mariotte into the 18th
        century. The three chapters on 18th-century topics
        deal respectively with Bouguer, Lambert and 18th-century
        work on phosphorescence.

(g) Heat

(i) General

599.   Bachelard, Gaston, *The Psychoanalysis of Fire*, trans.
        Alan C.M. Cross. London: Routledge & Kegan Paul,
        1964. x + 115 pp.

        First published in French in 1938. Includes a chapter
        on "The Chemistry of Fire: History of a False Problem"
        that deals with 18th-century theories of fire, especially
        Boerhaave's. Sees the problem, however, as really not
        one of scientific history but rather of "the history of
        the confusions that have been accumulated in the field
        of science by intuitions about fire."

600. Badcock, A.W., "Physics at the Royal Society, 1660-1800:
      I. Change of State." *Ann. Sci.* 16 (1960): 95-115.

      Briefly describes the work of a large number of
      scientists on aspects of the change of state problem,
      using the Society's *Journal Book* as the principal
      source.

601. Bellone, Enrico, "Osservazioni su alcuni aspetti della
      termologia del settecento, con particolare riferimento
      alle esperienze di Benjamin Thompson." *Physis* 13
      (1971): 376-398.

      Suggests that theories of heat were not necessarily
      rigidly separated as either fluid models or theories
      involving matter in motion, looking especially at
      Thompson's ideas and experiments.

602. Brown, Sanborn C., "The Discovery of Convection Currents
      by Benjamin Thompson, Count Rumford." *Amer. J. Phys.*
      15 (1947): 273-274.

      Largely consists of quotation from Rumford's 1797 paper
      wherein he describes his accidental discovery of con-
      vection currents and his subsequent experimental inves-
      tigations of them.

603. Brown, Sanborn C., "Count Rumford's Concept of Heat."
      Amer. J. Phys. 20 (1952): 331-334.

      Argues that Rumford's model for heat in 1800 was based
      on a close analogy with sound, implying that heat was
      a regular vibrational motion in the "fibres of metals"
      or through the aether and that temperature was related
      to frequency.

604. Brown, Sanborn C., *Benjamin Thompson, Count Rumford:*
      Count Rumford on the Nature of Heat. Oxford: Pergamon
      Press, 1967. ix + 207 pp.

      Contains reprints of, or extracts from, Rumford's
      eight most important papers on the vibrational theory
      of heat, together with a brief commentary, a biographical
      sketch and three previously published papers by Brown.
      Robert Fox (*Brit. J. Hist. Sci.* 4 (1968-69): 290-291)
      points out that the opposing contemporary caloric theory
      in the form described by Brown in fact postdates Rum-
      ford's death by three years and includes features which
      would have been unfamiliar to him.

*   Brown, Sanborn C., *Benjamin Thompson, Count Rumford*.

       Cited herein as item 38.

605. Cohen, I. Bernard, "Franklin, Newton, Boerhaave, Boyle and the Absorption of Heat in Relation to Colors." *Isis* 46 (1955): 99–104.

       Argues that Franklin's experiment on the heat absorption of various colours can probably be traced to his reading of Boerhaave, who in turn is likely to have derived the idea from Newton and Boyle.

606. Cornell, E.S., "Early Studies in Radiant Heat." *Ann. Sci.* 1 (1936): 217–225.

       Describes experiments and theories from the latter part of the 18th century due to Scheele, de Saussure, Pictet and Prévost in which it was generally agreed that radiant heat existed and behaved similarly to, but was distinct from, light. Describes some theories of the nature of heat, both material and vibratory.

607. Cornell, E.S., "The Radiant Heat Spectrum from Herschel to Melloni. I: The Work of Herschel and his Contemporaries." *Ann. Sci.* 3 (1938): 119–137.

       Describes Herschel's initial discovery that led him to the conclusion that there was a heating agency beyond the red part of the spectrum that seemed to experience refraction and polarization, and the reception of this idea, 1800–1830. For the continuation of the story, see item 956.

608. Donovan, A.L., *Philosophical Chemistry in the Scottish Enlightenment: The Doctrines and Discoveries of William Cullen and Joseph Black*. Edinburgh: at the University Press, 1975. x + 343 pp.

       Includes extended discussions of the views of Cullen and especially Black on fire and heat, and of the influence of Black's work on Watt, Irvine and others.

608a. Duveen, Denis I., and Roger Hahn, "Deux lettres de Laplace à Lavoisier." *Rev. Hist. Sci.* 11 (1958): 337–342.

       Publishes, with commentary, two letters from the early 1780s relating to Laplace's famous collaboration with Lavoisier in experiments on heat.

609. Dyck, David Ralph, "The Nature of Heat and its Relation-
      ship to Chemistry in the 18th Century." Ph.D. dis-
      sertation, University of Wisconsin, 1967.   235 pp.

      Firstly, has three chapters briefly describing the
      dynamic theory of heat (Hales, Desaguliers, Daniel
      Bernoulli), evolved for physical problems and out of
      favor by mid-18th century; the essentially "chemical"
      material theory of heat deriving from phlogiston theory
      (Rouelle, Macquer, Eller, culminating in Lavoisier);
      and a sample of a hybrid of the previous two, Boerhaave's
      vibrating fire particles.   Then turns to Black's work
      with his profound distinction between intensity and
      quantity of heat, and the physical strengths and chemical
      weakness of his theory.   Concludes briefly by taking
      the story to the end of the century when, it is argued,
      British and French theories had essentially drawn to-
      gether.

*   Fox, Robert, "The Science of Fire: J.H. Lambert and
      Study of Heat."

      Cited herein in item 418.

610. Gibbs, F.W., "Boerhaave's Chemical Writings." *Ambix* 6
      (1957-58): 117-135.

      A descriptive bibliography.

611. Goldfarb, Stephen J., "Rumford's Theory of Heat: A Re-
      assessment." *Brit. J. Hist. Sci.* 10 (1977): 25-36.

      Shows the centrality of motion and an all-pervasive
      aether in Rumford's natural philosophy and his rejection
      of unexplained forces.   Argues that his theory of heat
      was based on Boerhaave's "quasi-material" fire and may
      anticipate the wave models of heat popular during the
      1830s and 1840s.

612. Guerlac, Henry, "Chemistry as a Branch of Physics: La-
      place's Collaboration with Lavoisier." *Hist. Studs.
      Phys. Sci.* 7 (1976): 193-276.

      Describes the Lavoisier-Laplace collaboration of 1777
      on the vaporization of fluids--involving theories of
      matter, heat and fire--and that of 1781-1784 on heat.
      Shows the increasing influence of Laplace in developing
      theories and experiments that gave Lavoisier a firm
      basis for his anti-phlogiston theories.   Includes a
      description of their collaboration with Volta in an

investigation of atmospheric electricity. Concludes
with an analysis of their respective contributions to
the "Mémoire sur la chaleur" of 1783, discussing, *inter
alia*, the development of the ice calorimeter, the refining
of the use of specific heats, Laplace's theorizing on
heat, and the influence of the experiments on Lavoisier's
later work.

613. Guerlac, Henry, *Lavoisier--The Crucial Year: The Background and Origins of his First Experiments on Combustion in 1772.* Ithaca, N.Y.: Cornell University Press/
Oxford: Oxford University Press, 1961. xx + 240 pp.

Drawing on original sources (printed in an appendix)
provides an account of both the theoretical and experimental developments by which Lavoisier achieved his influential theory of combustion in 1772, examining,
inter alia, the role of the concept of elemental fire
in that achievement. Discusses Lavoisier's work in the
context of that of his contemporaries, both French and
English, arguing that the importance of the latter (especially Cullen and Black) has been overstated. Well
documented.

613a. Lavoisier, Antoine Laurent, and Pierre Simon Laplace,
*Memoirs on Heat, read to the Royal Academy of Sciences,
28 June 1783, by Messrs. Lavoisier and De La Place
of the Same Academy.* Translated with an introduction
by Henry Guerlac. New York: Neale Watson Academic
Publications, 1982. xviii + 110 pp.

Not seen.

614. Lodwig, T.H., and W.A. Smeaton, "The Ice Calorimeter of
Lavoisier and Laplace and some of its Critics."
Ann. Sci. 31 (1974): 1-18.

The authors describe the criticisms of the Lavoisier-
Laplace calorimeter raised by contemporaries and subse-
quently--by Wedgwood, Nicholson, Thomas Thomson, John
Herschel, Bunsen, etc.--and show that the apparatus
was used only infrequently in research.

615. Love, Rosaleen, "Some Sources of Herman Boerhaave's
Concept of Fire." *Ambix* 19 (1972): 157-174.

Describes Boerhaave's influential concept of a uni-
versally distributed material fire, continually in
motion and acting both through motion and by counter-
balancing the universal attractive forces that acted

between particles of ordinary matter. Finds sources
for these ideas in many medieval, Renaissance and 17th-
century authors.

616. Love, Rosaleen, "Herman Boerhaave and the Element-Instru-
      ment Concept of Fire." *Ann. Sci.* 31 (1974): 547-559.

      Shows that for Boerhaave fire was both a unique kind
      of matter and an unchanging instrument of chemical change.
      Discusses some influential antecedent and contemporary
      views on the subject.

617. Lovell, D.J., "Herschel's Dilemma in the Interpretation
      of Thermal Radiation." *Isis* 59 (1968): 46-60.

      Gives an account of William Herschel's discovery of
      spectral heat, and how he rejected its identity with
      light on the basis of his poor data and its seeming
      impossibility from a Newtonian viewpoint.

618. McKie, Douglas, and N.H. de V. Heathcote, *The Discovery
      of Specific and Latent Heats*. London: Edward Arnold
      & Co., 1935. 155 pp.

      A detailed and very thorough account which focuses on
      the work of Black and Wilcke in particular.

619. McKie, Douglas, and N.H. de V. Heathcote, "William Cleg-
      horn's *De Igne* (1779)." *Ann. Sci.* 14 (1958): 1-82.

      A translation and reprint, with extensive annotations,
      of Cleghorn's M.D. dissertation, the first detailed
      exposition of the material theory of heat.

 *   Metzger, Hélène, *Newton, Stahl, Boerhaave et la doctrine
      chimique*.

      Cited herein as item 430. Includes an extensive dis-
      cussion of Boerhaave's theory of elemental fire.

620. Middleton, W.E. Knowles, "Jacob Hermann and the Kinetic
      Theory." *Brit. J. Hist. Sci.* 2 (1965-66): 247-250.

      A translation of a brief early (1716) treatment of
      a gas from a kinetic viewpoint.

621. Morris, Robert J., "Eighteenth-Century Theories of the
      Nature of Heat." Ph.D. dissertation, University of
      Oklahoma, 1965.

      Describes the theories: the basically English view
      that heat was a vibrating motion; the more commonly

accepted theory espoused by Boerhaave that heat was
fire particles in motion; and the increasingly common
view derived from Lavoisier from the 1770s, that heat
was present, as caloric, in gases. Concludes with a
discussion of criticisms of the caloric theory.

622. Morris, Robert J., "Lavoisier and the Caloric Theory."
    Brit. J. Hist. Sci. 6 (1972-73): 1-38.

    Traces Lavoisier's development of the caloric theory
    from a purely chemical to a more physical approach,
    ultimately based on a balance-of-forces concept that
    explained a great variety of phenomena.

623. Pacey, A.J., "Some Early Heat Engine Concepts and the
    Conservation of Heat." *Brit. J. Hist. Sci.* 7 (1974):
    135-145.

    Argues that in the 1780s, Lee, Ewart and Watt all had
    the idea of conservation of heat in heat flow, and that
    Smeaton, Lee and Ewart also conjectured that there was
    a relationship between heat, temperature, and mechanical
    power.

624. Partington, J.R., and Douglas McKie, "Historical Studies
    on the Phlogiston Theory--"I. The Levity of Phlogiston";
    "II. The Negative Weight of Phlogiston"; "III. Light
    and Heat in Combustion"; "IV. Last Phases of the Theory."
    Ann. Sci. 2 (1937): 361-404; 3 (1938): 1-58, 337-371;
    4 (1939-40): 113-149.

    While mainly concerned with chemical aspects of the
    phlogiston theory, these papers also detail a number
    of opinions on the interrelationships between heat,
    light, phlogiston and elemental fire from the last third
    of the eighteenth century.

*   Robinson, Eric, and Douglas McKie, eds., *Partners in
    Science: Letters of James Watt and Joseph Black*.

    Cited herein as item 444.

624a. Scott, E.L., "Richard Kirwan, J.H. de Magellan, and the
    Early History of Specific Heat." *Ann. Sci.* 38 (1981):
    141-153.

    Discusses Magellan's attribution to Kirwan of the
    table of specific heats that he published in 1780--the
    first such table to appear in print--and his correspon-
    dence with James Watt over Joseph Black's role in the
    discovery of specific heat capacities.

624b. Sebastiani, Fabio, "La memoria Voltiana intorno al
       calore." *Physis* 23 (1981): 89-113.

       An analysis of Volta's ideas on heat which draws atten-
       tion to his indebtedness to the work of William Irvine.

625.   Talbot, G.R., and A.J. Pacey, "Antecedents of Thermo-
       dynamics in the Work of Guillaume Amontons." *Centaurus*
       16 (1971): 20-40.

       The authors discuss Amontons' work (mainly 1699-1702)
       on temperature scales and the notion of absolute zero,
       on gas physics, and on heat machines, and also his in-
       fluence.

626.   Talbot, G.R., and A.J. Pacey, "Some Early Kinetic Theories
       of Gases: Herapath and His Predecessors." *Brit. J.
       Hist. Sci.* 3 (1966): 133-149.

       The authors summarize several speculative theories from
       the 18th century including those of Jean Bernoulli and
       Euler which invoked rotating particles, Daniel Bernoulli's
       which invoked translational motion, and Lomonosov's
       which invoked both. They also briefly discuss Herapath's
       theory, ascribing its weaknesses to the contradiction
       between his use of "Newtonian" hard particles and his
       need for (but rejection of) *vis viva*.

627.   Watanabe, Masao, "Count Rumford's First Exposition of
       the Dynamic Aspect of Heat." *Isis* 50 (1959): 141-144.

       Shows that Rumford gave an early, less detailed, ac-
       count of his theory of heat in his discussion of his
       experiments on gunpowder (begun 1778, published 1781),
       wherein he concluded that the heating effects of a gun
       discharge were due to a vibration brought about by the
       expansion and contraction of the barrel.

628.   Watanabe, Masao, "James Hutton's '*Obscure* Light': A Dis-
       covery of Infrared Radiation predating Herschel's."
       Japanese Studs. Hist. Sci. 17 (1978): 97-104.

       Argues that the "obscure light" whose existence was
       announced by Hutton in 1794 was in fact infrared radia-
       tion. Hutton so entangled his discovery, however, with
       his "incomprehensible" theories of light and heat, that
       its significance was overlooked.

(ii) Thermometry

629. Austin, Jillian F., and Anita McConnell, "James Six,
      F.R.S.--Two Hundred Years of the Six's Self-Registering
      Thermometer." *Notes Rec. Roy. Soc. Lond.* 35 (1980):
      49-65.

      Biography of Six including an account of his thermo-
      metric writings, followed by a description of the in-
      vention and development of his thermometer and details
      of twenty-six surviving examples of Six's instruments.

630. Austin, Jill, and Anita McConnell, eds., *The Construction
      and Use of a Thermometer, by James Six, F.R.S., Pre-
      faced by an Account of his Life and Works and the Use
      of his Thermometer over Two Hundred Years.* London:
      Nimbus Books, 1980. 123 pp.

      A photoreproduction of Six's posthumously published
      book describing his maximum-and-minimum thermometer,
      together with sundry items concerning Six's life and
      work and the uses to which his instrument was subse-
      quently put.

631. Birembaut, Arthur, "La contribution de Réaumur à la
      thermométrie." *Rev. Hist. Sci.* 11 (1958): 302-329.

      Describes the problems thermometry faced in Réaumur's
      day, details the modifications he introduced including
      his temperature scale, and discusses the criticisms of
      his work from 1730 to the 1790s.

632. Chaldecott, J.A., "Cromwell Mortimer, F.R.S. (*c.* 1698-
      1752), and the Invention of the Metalline Thermometer
      for Measuring High Temperatures." *Notes Rec. Roy.
      Soc. Lond.* 24 (1969): 113-135.

      Describes Mortimer's 1735 design based on the expansion
      properties of metals when heated. Argues that the ther-
      mometer was never fully calibrated, and that the frequent
      citations of his values of metallic melting points
      during the eighteenth century were based on arbitrary
      values which he had inscribed on a diagram of his ther-
      mometer. These values were accepted until new work was
      done in the field at the end of the century.

633. Chaldecott, J.A., "Hans Loeser's Metallic Thermometers
      of 1746 and 1747." *Ann. Sci.* 28 (1972): 87-100.

      Describes Loeser's thermometers from near-contemporary
      descriptions and surviving models. Argues that they were

independent of Mortimer's earlier apparatus and, while
better, were almost totally unknown in the eighteenth
and early nineteenth centuries.

634. Chaldecott, J.A., "The Platinum Pyrometers of Louis
      Bernard Guyton de Morveau, F.R.S. (1737-1816)." *Ann.
      Sci.* 28 (1972): 347-368.

      Describes from MS sources the development of Guyton's
      high temperature thermometer during the period 1803-10,
      comparing this with Wedgwood's much older and inferior
      apparatus.

635. Cohen, Ernst, and W.A.T. Cohen-De Meester, "Daniel Gabriel
      Fahrenheit." *Verhand. Nederlandsche Akad. Weten.:
      Afdeeling Natuurkunde, Sect. I* 16(2) (1936). 37 pp.

      Draws on correspondence between Fahrenheit and Boerhaave
      to uncover some details of Fahrenheit's life when he was
      deriving his thermometric scale. Maintains that the
      scale was derived from that of Roemer. Includes a brief
      discussion of Fahrenheit's lecture courses during that
      time.

636. Cohen, I. Bernard, "Roemer and Fahrenheit." *Isis* 39
      (1948): 56-58.

      Concludes that Fahrenheit gained little from Roemer
      in developing his thermometer scale.

*   Meyer, Kirstine, *Die Entwicklung des Temperaturbegriffes
      im Laufe der Zeiten.*

      Cited herein as item 335.

*   Middleton, W.E. Knowles, *A History of the Thermometer
      and Its Use in Meteorology.*

      Cited herein as item 298.

                    (h) Electricity and Magnetism

637. *Aepinus's Essay on the Theory of Electricity and Mag-
      netism.* Introductory monograph and notes by R.W.
      Home. Translation by P.J. Connor. Princeton, N.J.:
      Princeton University Press, 1979. xiv + 514 pp.

Includes a full English translation of Aepinus' *Essay* (1759), together with an annotated bibliography of his published writings and a 224-page introductory monograph comprising a biographical outline and an extended discussion of the eighteenth-century sciences of electricity and magnetism and the place of Aepinus' work therein. It is argued that even though Aepinus was not prepared to assume any particular form for the laws of electrical and magnetic action, his work nevertheless constitutes the beginning of the mathematical sciences of electricity and magnetism, and thereby an important step in the mathematization of physics more generally.

637a. [Aepinus, F.U.T.]. *Rostocker Wissenschaftshistorische Manuskripte, Heft 3.* Rostock: Wilhelm-Pieck-Universität, Sektion Geschichte, 1979. 101 pp.

Includes the following papers presented at a 1978 colloquium devoted to the life and work of Aepinus: Martin Guntau, "Zu den deutsch-russischen Wissenschaftsbeziehungen im 18. Jahrhundert und F.U.Th. Aepinus"; Dieter Hoffmann, "Franz Ulrich Theodor Aepinus--ein beinahe vergessener Pionier der Elektrizitätslehre"; Jürgen Hamel, "Der Beitrag von F.U.Th. Aepinus zur Erforschung des Erdmondes"; Bernhard Wandt, "Zur Geschichte der Familie Aepinus an der Universität Rostock"; Annette Vogt, "Zur Einheit von wissenschaftlicher und wissenschaftsorganisatorischer Tätigkeit im Wirken von Leonhard Euler."

637b. Benguigui, Isaac, "La théorie d'électricité de Nollet et son application en médecine à travers sa correspondance inédite avec Jallabert." *Gesnerus* 38 (1981): 225-235.

Draws attention to Nollet's interesting and extensive correspondence with Jallabert, highlighting remarks by Nollet about the status of his electrical theory and exchanges between the two men concerning possible medical applications of electricity.

638. Benz, Ernst, *Theologie der Elektrizität: Zur Begegnung und Auseinandersetzung von Theologie und Naturwissenschaft im 17. und 18. Jahrhundert.* Mainz: Akademie der Wissenschaft und der Literatur, 1970. 98 pp.

Discusses the way in which, for some 17th and 18th-century thinkers, electricity (and also magnetism) came

to be seen as a new symbol of God, a sensible manifesta-
tion of the hidden presence of God's power in the world,
in the way that light had been seen in the Middle Ages.
The chief 17th-century authors discussed are Rudolf
Goclenius and Athanasius Kircher, while from the 18th
century, those whose work is examined are Franklin,
Mesmer, and especially Friedrich Christoph Oetinger,
Johann Ludwig Fricker and Prokop Diviš.

*   Brunet, Pierre, "L'oeuvre scientifique de Charles
      François Du Fay (1698-1739)."

      Cited herein as item 360.

639. Brunet, Pierre, "Les origines du paratonnerre." *Rev.
      Hist. Sci.* 1 (1948): 213-253.

      Describes Franklin's theory and his initial suggestion
      of the lightning rod, Nollet's opposition to it, and
      contributions to the debate by Dalibard, Beccaria, Le
      Roy, Wilson and others, concluding with an account of
      the gradual introduction of the Franklinist rod in
      various parts of Europe.

640. Brunet, Pierre, "Les premières recherches expérimentales
      sur la foudre et l'électricité atmosphérique."
      Lychnos 10 (1946-47): 117-148.

      Surveys the evolution of the electrical theory of
      lightning during the period 1746-1752, discussing the
      work of Franklin, Nollet, Dalibard and Le Monnier amongst
      many others. A good primary bibliography.

640a. Chenakal, V.L., "Elektricheskie mashiny v Rossii XVIII
      veka" ("Electrical machines in 18th-century Russia").
      Istoriya fiziko-matematicheskikh nauk 43 (1961): 50-
      111.

      A detailed survey of surviving machines in Russian
      museum collections that also incorporates a considerable
      amount of valuable related material from the archives
      of the Soviet Academy of Sciences.

641. Chipman, R.A., "An Unpublished Letter of Stephen Gray
      on Electrical Experiments, 1707-1708." *Isis* 45
      (1954): 33-40.

      Publishes a letter from Gray to Hans Sloane on elec-
      trical experiments. Suggests that the non-publication
      of the letter in the *Philosophical Transactions* may have

been due to Flamsteed's sponsorship of Gray and the
antipathy of Newton and Sloane to Flamsteed.

642. Chipman, R.A., "The Manuscript Letters of Stephen Gray,
      F.R.S., 1666/7-1736." *Isis* 49 (1958): 414-433.

      Lists the more than fifty MS letters of Gray known
      to exist and uses them to expand the limited biographical
      information on him, providing new information on his
      scientific interests and contacts.

642a. Clark, David H., and Lesley Murdin, "The Enigma of Stephen
      Gray, Astronomer and Scientist (1666-1736)." *Vistas in
      Astronomy* 23 (1979): 351-404.

      Includes a considerable amount of new biographical infor-
      mation about the pioneering electrical investigator, espe-
      cially concerning his early astronomical work.

643. Cohen, I. Bernard, "Guericke and Dufay." *Ann. Sci.* 7
      (1951): 207-209.

      Shows that Dufay in 1733 was one of the first to
      present Guericke as a contributor to electrical science.

644. Cohen, I. Bernard, "Neglected Sources for the Life of
      Stephen Gray (1666 or 1667-1736)." *Isis* 45 (1954):
      41-50.

      Argues for an earlier birthdate (1666 or 1667) than
      previously assigned to Gray. Shows that Gray for a
      time was at Trinity College, Cambridge, as assistant
      to Cotes in the observatory then under construction
      (*c.* 1707-1708), and that he lived for a time with J.T.
      Desaguliers in London as his assistant. Concludes with
      a bibliography of Gray's published papers.

*   Cohen, I. Bernard, *Franklin and Newton.*

      Cited herein as item 370.

645. Cohen, I. Bernard, and Robert Schofield, "Did Diviš
      Erect the First European Protective Lightning Rod,
      and Was His Invention Independent?" *Isis* 43 (1952):
      358-364.

      It is argued that Diviš appears to have been conversant
      with much contemporary work on electricity and was thus
      unlikely to have been totally ignorant of Franklin's
      work, and, further, that the protective rod was already
      known in Europe at the time he did his work. Cf. item
      685 for a contrary view.

646. Corson, D.W., "Pierre Polinière, Francis Hauksbee and
      Electroluminescence: A Case of Simultaneous Discovery."
      Isis 59 (1968): 402-413.

      Shows that Hauksbee, an established investigator, was
      the generator of further work, while the popularizer
      Polinière neither pursued his virtually simultaneous
      discoveries nor influenced anyone else to do so.

647. Daujat, Jean, *Origines et formation de la théorie des
      phénomènes électriques et magnétiques.* 3 vols.
      Paris: Hermann, 1945. (Actualités scientifiques et
      industrielles, 989-991).

      The three volumes deal respectively with "Antiquité
      et Moyen Age," "Le XVIIe siècle," and "Le XVIIIe
      siècle." The style is the same throughout: the final
      volume comprises brief uncritical descriptions of the
      ideas of each of the leading contributors to 18th-
      century electricity and magnetism, and also of some of
      the lesser lights such as Hartsoeker, Du Tour, Jallabert
      and Paulian.

648. Dibner, Bern, "The Great Van Marum Electrical Machine."
      Natural Philosopher 2 (1963): 89-103.

      Describes Van Marum's machine and the experiments he
      performed with it.

649. Dorling, Jon, "Henry Cavendish's Deduction of the
      Electrostatic Inverse Square Law from the Result of a
      Single Experiment." *Studs. Hist. Phil. Sci.* 4 (1974):
      327-348.

      Argues that Cavendish was able to deduce the law
      logically from one negative experiment, given only the
      inherently "implausible" assumption that electrostatic
      force varies as an inverse power of the distance; but
      further concludes that, given the contemporary theoretical
      and experimental background, the assumption was reasonable
      at the time.

650. Espenschied, Lloyd, "More on Franklin's Introduction to
      Electricity." *Isis* 46 (1955): 280-281.

      Argues that in between Franklin's first two encounters
      with electrical phenomena as mentioned in N.H. De V.
      Heathcote's article (item 669) electricity "had become
      the centre of attraction in intellectual and aristocratic
      circles" and frequently mentioned in print. *cf.* L. Es-
      penschied, "The Electrical Flare of the 1740s." *Elec-
      trical Engineering* 74 (1955): 392-397.

651. Finn, Bernard S., "The Influence of Experimental Apparatus on Eighteenth-Century Electrical Theory." *Proceedings, 12th International Congress of the History of Science, Paris, 1968* (Paris, 1971), vol. 10A, pp. 51-55.

Gives examples of the dependence of particular electrical phenomena, and hence particular theories, on the nature of the apparatus used.

652. Finn, Bernard S., "An Appraisal of the Origins of Franklin's Electrical Theory." *Isis* 60 (1969): 362-369.

Argues that Franklin was ignorant of most European electrical theories, and that this isolation allowed him to ignore problems that plagued European theorists and to strike out in a new direction based almost entirely on William Watson's work and on newly discovered phenomena such as the Leyden jar. The argument is controverted, at least in part, by Heilbron's more recent work (item 674).

653. Finn, Bernard S., "Output of Eighteenth-Century Electrical Machines." *Brit. J. Hist. Sci.* 5 (1970-71): 289-291.

Provides tables showing the output of various machines, and suggests that the Leyden jar would have revolutionized electrical studies by increasing the available output "by several orders of magnitude." A different assessment of output is given by H. Prinz (item 695).

654. Fisher, J., "The History of Electricity in Germany during the First Half of the Eighteenth Century." *Bull. Brit. Soc. Hist. Sci.* 2 (1956-57): 49-51.

A brief survey of early German work on electricity, very welcome at the time for drawing the attention of English-language historians to the importance of the German contributions, but now superseded by Heilbron's much more extensive treatment of the subject (in item 400).

655. [Franklin, Benjamin], *Benjamin Franklin's Experiments: A New Edition of Franklin's Experiments and Observations on Electricity*, ed. with a critical and historical introduction by I.B. Cohen. Cambridge, Mass.: Harvard University Press, 1941. xxviii + 453 pp.

The definitive edition of Franklin's papers on electricity. The historical introduction is now, however, rather dated.

656. Freudenthal, Gad, "Early Electricity between Chemistry
      and Physics: The Simultaneous Itineraries of Francis
      Hauksbee, Samuel Wall, and Pierre Polinière." *Hist.
      Studs. Phys. Sci.* 11 (1981): 203-229.

      Argues that it was no coincidence that several early
      18th-century investigators moved from studying the
      "mercurial phosphorus" to the study of electricity;
      rather, they were led in this direction by current
      theories concerning the origin of the light emitted by
      phosphors more generally.

657. Gill, Sydney, "A Voltaic Enigma and a Possible Solution
      to It." *Ann. Sci.* 33 (1976): 351-370.

      Suggests that although Volta had all the apparatus
      required for his "pile" by 1792, he did not construct
      one for another eight years because he was more concerned
      with refuting the attacks upon his theory by Galvani
      and his supporters.

*   Gillmor, C. Stewart, *Coulomb and the Evolution of Physics
      and Engineering in Eighteenth-Century France.*

      Cited herein as item 733.

658. Gliozzi, Mario, "Studio comparativo delle teorie elet-
      triche del Nollet, del Watson e del Franklin."
      Archeion 15 (1933): 202-215.

      Argues that Watson in effect proposed two theories
      of electricity, one being similar to that of Nollet
      and the other to that of Franklin.

659. Gliozzi, Mario, "Giambatista Beccaria nella storia
      dell'elettricità." *Archeion* 17 (1935): 15-47.

      Describes Beccaria's work on electricity and mag-
      netism, showing its influence on Volta and the simi-
      larity of his ideas to those of Faraday.

660. Gliozzi, Mario, *L'elettrologia fino al Volta.* 2 vols.
      Naples, 1937.

      An excellent survey, unfortunately very rare as a
      result of war-time destruction of the publisher's stocks.

*   Gliozzi, Mario, *Fisici piemontesi del Settecento nel
      movimento filosofico del tempo.*

      Cited herein as item 382. Concerned chiefly with
      Beccaria.

661. Gliozzi, Mario, "Il Volta della seconda maniera." *Cultura e scuola* 5 (1966): 235-239.

   Draws attention to a marked change in Volta's thinking about electricity that occurred in the late 1770s, when he abandoned the old-style electrical atmospheres in favor of forces acting at a distance.

662. Gliozzi, Mario, "Consonanze e dissonanze tra l'elettrostatica di Cavendish e quella di Volta." *Physis* 11 (1969): 231-248.

   Suggests that Cavendish's papers of 1771 and 1776 may have influenced Volta and compares their respective conceptual and experimental treatments of similar problems.

663. Gross, B., "On the Experiment of the Dissectible Condenser." *Amer. J. Phys.* 12 (1944): 324-329.

   While mostly concerned with various 20th-century explanations of the phenomenon, Gross describes Franklin's initial experiment with the dissectible condenser and his conclusion that the "accumulation of the electrical fire be in the electrified glass."

664. Guerlac, Henry, "Francis Hauskbee: expérimentateur au profit de Newton." *Arch. Int. Hist. Sci.* 16 (1963): 113-128. Reprinted, *Essays and Papers in the History of Modern Science*, by Henry Guerlac. Baltimore and London: Johns Hopkins University Press, 1977, pp. 107-119.

   Argues that Hauksbee's experiments on capillarity and electricity, about 1705, had a strong influence on Newton's thinking and contributed to his revival of an ether theory.

665. Hackmann, Willem D., "The Design of the Triboelectric Generators of Martinus van Marum, F.R.S.: A Case History of the Interaction between England and Holland in the Field of Instrument Design in the Eighteenth Century." *Notes Rec. Roy. Soc. Lond.* 26 (1971): 163-181.

   By tracing the design of Van Marum's generators, shows the strong English influence on Dutch practice, whether directly through the aid of English instrument makers such as Cuthbertson or through advice solicited in correspondence, or less directly such as through Van Marum's study of current English research. Additionally,

the extensive correspondence maintained by Van Marum
meant that the English could draw freely on the results
of his own investigations and could imitate his much
admired designs.

665a. Hackmann, Willem D., "The Researches of Dr. Martinus
       Van Marum (1750-1837) on the Influence of Electricity
       on Animals and Plants." *Medical History* 16 (1972):
       11-26.

       Briefly surveys a number of 18th-century theories
       concerning the role of electricity in living organisms
       before describing Van Marum's work.

666.   Hackmann, Willem D., *John and Jonathan Cuthbertson:
       The Invention and Development of the Eighteenth-
       Century Plate Electrical Machine*. Leiden: Rijks-
       museum voor de Geschiedenis der Natuurwetenschappen,
       1973. 72 pp.

       Contains biographical information about the Cuth-
       bertson brothers, a brief account of the invention of
       the plate generating machine, and a description of the
       various innovations introduced by John Cuthbertson to
       make the best plate glass generators.

667.   Hackmann, Willem D., *Electricity from Glass: The History
       of the Frictional Electrical Machine, 1600-1850*.
       Alphen aan den Rijn: Sijthoff and Noordhoff, 1978.
       xiii + 310 pp.

       An excellent account of the changes and developments
       in generator design in the period, which the author
       attempts to link with developments in the general history
       of electrical theory. Reviewed by J.L. Heilbron, *Ann.
       Sci.* 37 (1980): 116-117, and R.W. Home, *Brit. J. Hist.
       Sci.* 13 (1980): 268-270.

668.   Hackmann, Willem D., "The Relationship between Concept
       and Instrument Design in Eighteenth-Century Experimental
       Science." *Ann. Sci.* 36 (1979): 205-224.

       Examines the mutual interaction between empirically
       based theory and the design of scientific instruments
       in 18th-century electrical science, the way this affected
       the generation and interpretation of the phenomena, and
       the limitations of the 18th-century approach.

669.   Heathcote, N.H. de V., "Franklin's Introduction to Elec-
       tricity." *Isis* 46 (1955): 29-35.

Argues that Franklin's first (limited) introduction
to electrical experiments was in the lectures he attended by Dr. Adam Spencer in 1743, but that his interest
developed only with Peter Collinson's gift to the Library
Company of Philadelphia of a glass tube and directions
for its use in 1746.

670. Heathcote, N.H. de V., "The Early Meaning of *Electricity*:
     some *Pseudodoxia Epidemica*." *Ann. Sci.* 23 (1967):
     261-275.

     Shows that throughout the 17th and most of the 18th
     century the word "electricity" meant "the power, property
     or quality of attracting light bodies" and said nothing
     about the *cause* of the attraction.

671. Heilbron, J.L., "A propos de l'invention de la bouteille
     de Leyde." *Rev. Hist. Sci.* 19 (1966): 133-142.

     Argues that as Musschenbroek was teaching and was thus
     familiar with contemporary electrical theory, and that
     this would have strongly discouraged him from trying
     the experimental arrangement required for the Leyden
     experiment to succeed, it is more likely to have been
     carried out by the dilettante Cunaeus.

672. Heilbron, J.L., "G.M. Bose: The Prime Mover in the In-
     vention of the Leyden Jar?" *Isis* 57 (1966): 264-267.

     Suggests that it was Musschenbroek's repeating and
     modifying an earlier experiment suggested by Bose that
     provided the opportunity for the unexpected discovery
     of the power of the Leyden jar.

673. Heilbron, J.L., "Robert Symmer and the Two Electricities."
     Isis 67 (1976): 7-20.

     Provides new biographical information concerning
     Symmer and an account of his two-fluid theory of elec-
     tricity, published in 1759, which proved immensely in-
     fluential amongst the continental "dualist" electricians.

674. Heilbron, J.L., "Franklin, Haller, and Franklinist His-
     tory." *Isis* 68 (1977): 539-549.

     Shows Franklin's debt at the outset of his electrical
     investigations to a brief account of recent German work
     written by Albrecht von Haller. Denounces the strong
     Franklinist bias of the leading English-language his-
     tories of electricity, arising initially from the ex-

cessive parochialism of Priestley's *History* but still
readily apparent, according to Heilbron, in current
writing on the subject.

*   Heilbron, John L., *Electricity in the 17th and 18th Cen-
        turies: A Study of Early Modern Physics.*

        Cited herein as item 400.

*   Heilbron, J.L., "The Electrical Field before Faraday."

        Cited herein in item 238.

674a.   Hindle, Brook, "David Rittenhouse's Theory of Magnetic
        Dipoles." *Proceedings, 10th International Congress
        of the History of Science, Ithaca, 1962* (Paris, 1964),
        Vol. 2, pp. 755-758.

        Briefly discusses Rittenhouse's theory of molecular
        magnets, stressing his wariness of hypotheses and his
        unwillingness to speculate about the ultimate cause of
        magnetism.

675.   Home, R.W., "Francis Hauksbee's Theory of Electricity."
        Arch. Hist. Exact Sci. 4 (1967-68): 203-217.

        Describes the interplay between Hauksbee's important
        electrical experiments and his elaboration of the then
        generally held notion that electrical phenomena were
        brought about by streaming matter of some kind.

676.   Home, R.W., "Aepinus and the British Electricians: The
        Dissemination of a Scientific Theory." *Isis* 63 (1972):
        190-204.

        Presents evidence that Aepinus' work was both known
        to and admired by British "electricians," but that it
        nevertheless had little impact on their thinking. Sug-
        gests reasons for this.

677.   Home, R.W., "Franklin's Electrical Atmospheres." *Brit.
        J. Hist. Sci.* 6 (1972-73): 131-151.

        Argues that Franklin's notion of an "atmosphere" of
        electric fluid surrounding an electrified body played a
        central role in his theory, but that it also gave rise
        to some awkward contradictions. These, together with
        Franklin's failure to explain adequately what he had in
        mind, led his followers to modify his theory in various
        ways before eventually adopting a thorough-going action-
        at-a-distance theory instead.

678. Home, R.W., "Electricity in France in the Post-Franklin
      Era." *Proceedings, 14th International Congress of the
      History of Science, Tokyo, 1974* (Tokyo, 1975), Vol. 2,
      pp. 269-272.

      Presents evidence to show that, at least in France,
      Franklin's theory of electricity did not immediately
      displace that of his rival Nollet following its publica-
      tion in the early 1750s, but that, on the contrary,
      Nollet's theory emerged victorious from the initial
      debate in the Paris Academy of Sciences and thereafter
      remained paramount until well into the 1770s.

679. Home, R.W., "Some Manuscripts on Electrical and other
      Subjects attributed to Thomas Bayes, F.R.S." *Notes
      Rec. Roy. Soc. Lond.* 29 (1974-75): 81-90.

      Describes some manuscripts attributed to Bayes (better
      known for his famous theorem about probabilities) and
      provides a decipherment of the system of shorthand used
      in these. Discusses Bayes' criticisms, as set out in
      one of these documents, of Benjamin Wilson's theory of
      electricity (1746, 1752, 1756).

680. Home, R.W., "On Two Supposed Works by Leonhard Euler
      on Electricity." *Arch. Int. Hist. Sci.* 25 (1975):
      3-7.

      Presents new evidence to show that two items published
      during the 1750s over the name of Leonhard Euler's son,
      Johann Albrecht Euler, were in fact written by him and
      not by the father as has been supposed by a number of
      recent authors.

681. Home, R.W., "Aepinus, The Tourmaline Crystal, and the
      Theory of Electricity and Magnetism." *Isis* 67
      (1976): 21-30.

      Describes Aepinus' investigation of the electrical
      properties of the tourmaline (1757), arguing that this
      provided important new support for Franklin's theory
      of electricity and also led Aepinus himself to his revo-
      lutionary new theory of magnetism.

682. Home, R.W., "'Newtonianism' and the Theory of the Mag-
      net." *Hist. Sci.* 15 (1977): 252-266.

      Shows that, like Descartes and his disciples, Newton
      and his followers in early 18th-century England believed
      that magnetic phenomena were brought about by streams

of subtle matter circulating through and around magnetized bodies. Argues from this that terms such as "Cartesian" and "Newtonian" must be used more cautiously than has sometimes been done.

683. Home, R.W., "Nollet and Boerhaave: A Note on Eighteenth-Century Ideas about Electricity and Fire." *Ann. Sci.* 36 (1979): 171-175.

Points out that although Nollet's theory of electricity eventually fell into disfavour, it enjoyed considerable support for a number of years in the mid-18th century. Argues that this was due, in part at least, to the theory's "intimate connections" with the theory of fire expounded by Boerhaave.

*   [Home, R.W.], *Aepinus's Essay on the Theory of Electricity and Magnetism*.

Cited herein as item 637.

684. Home, R.W., *The Effluvial Theory of Electricity*. New York: Arno Press, 1981. xxviii + 297 pp.

A hard-cover version of the author's 1967 Indiana University doctoral thesis, with a supplementary essay that takes account of more recent work. Argues that the notion that electrical effects were brought about by streams of subtle effluvia formed the basis of a coherent theoretical tradition during the first half of the 18th century. Includes extended discussions of the work of, *inter alia*, Hauksbee, Dufay, Nollet, Watson, Franklin and Beccaria, and gives a detailed account of the important debates on electrical theory that took place in the Paris Academy of Sciences during the 1750s.

*   Home, R.W., "Physical Principles and the Possibility of a Mathematical Science of Electricity and Magnetism."

Cited herein in item 742.

685. Hujer, Karel, "Father Procopius Diviš, the European Franklin." *Isis* 43 (1952): 351-357.

Argues that Diviš independently erected the first grounded lightning rod in Europe in 1756. Further describes some of his other experiments with electricity and the problems he had in publishing his work due to the repressive religious atmosphere in Moravia at the time. Cf. item 645 for a contrary view.

686. Leicester, Henry M., "The Electrical Theories of M.V.
        Lomonosov." *Ann. Sci*. 30 (1973): 299-310.

        Describes Lomonosov's theories based on his meteoro-
        logical studies and his belief in links between friction,
        heat and electricity. Relates various criticisms of
        the theory and Lomonosov's replies, but makes no attempt
        either to assess their adequacy or to place Lomonosov's
        work within the general context of 18th-century ideas
        about electricity.

687. Lemay, Joseph A. Leo, "Franklin and Kinnersley." *Isis*
        52 (1961): 575-581.

        Argues that Franklin's presumed earliest article on
        electricity is in fact an outline of the lectures given
        by his friend and co-researcher Ebenezer Kinnersley.
        Presents indications of a "well-accepted minority belief"
        in colonial America that Franklin plagiarized from Kin-
        nersley.

688. Lemay, Joseph A. Leo, *Ebenezer Kinnersley, Franklin's
        Friend*. Philadelphia: University of Pennsylvania
        Press, 1964. 143 pp.

        Describes the life and teachings of Kinnersley (1711-
        1778), who was responsible for the first popular lec-
        tures and research on electricity in America.

*   Loria, Mario, "Un manuscrit de l'Académie des Sciences
        de Turin: le *Traité de physique* de Jean-Baptiste
        Beccaria."

        Cited herein as item 422a.

689. McCormmach, Russell, "The Electrical Researches of
        Henry Cavendish." Ph.D thesis, Case Western Reserve
        University, 1967. 517 pp.

        A comprehensive analysis of Cavendish's electrical
        investigations, based on a thorough study of Cavendish's
        manuscripts as well as on Maxwell's published edition
        of the electrical papers.

690. McEvoy, John G., "Electricity, Knowledge and the Nature
        of Progress in Priestley's Thought." *Brit. J. Hist.
        Sci*. 12 (1979): 1-30.

        Argues that Priestley's work and writings on electricity
        must be understood as a deliberate application of his
        general epistemology and methodology.

691.   Pace, Antonio, *Benjamin Franklin and Italy*. Philadel-
       phia: American Philosophical Society, 1958. xvi +
       450 pp.

       Chiefly concerned with Franklin's connections with and
       influence in Italy as a statesman and popular philosopher
       and in literature and the arts. Also, however, includes
       valuable chapters on the general Italian response to
       his work on electricity and on the work of Beccaria in
       particular. An appendix gives the text of a number of
       letters, many previously unpublished, that bear upon
       these matters.

692.   Palter, Robert, "Early Measurements of Magnetic Force."
       Isis 63 (1972): 544-558.

       Describes a number of early attempts to find a law of
       magnetic force, including those of Newton, Hauksbee,
       Brook Taylor, Musschenbroek and John Michell. Draws
       attention to (and provides an English translation of)
       a long footnote in the Le Seur-Jacquier edition (1739-
       42) of Newton's *Principia* dealing with this subject,
       arguing that it was due to J.L. Calandrini.

*   Pater, Cornelis de, *Petrus van Musschenbroek (1692-1761),
       een newtoniaans natuuronderzoeker*.

       Cited herein as item 437. Includes a detailed dis-
       cussion of Musschenbroek's influential work on magnetism.

693.   Polvani, Giovanni, *Alessandro Volta*. Pisa, 1942.

       The best general account of Volta's work, but suffers
       from being relatively uncritical.

694.   Priestley, Joseph, *The History and Present State of Elec-
       tricity, with Original Experiments*. 2 vols. London,
       1767. Reprint of the 3rd edition (1775), New York:
       Johnson Reprint, 1966, with introduction by Robert E.
       Schofield.

       The starting point of all subsequent writings on the
       history of 18th-century electricity. Only in recent
       years has the extent of its Franklinist bias begun to
       be fully appreciated.

695.   Prinz, H., "Erschütterndes und Faszinierendes über
       gespeicherte Elektrizität." *Bull. Schweiz. Elektro-
       tech. Ver.* 62 (1971): 97-109.

Gives assessments of the output of 18th-century electrical machines which differ markedly from those given
by B.S. Finn (item 653).

696. Robinson, Myron, "A History of the Electric Wind."
Amer. J. Phys. 30 (1962): 366-372.

Provides a list of eighteenth-century references to
the electric wind (including those of Hauksbee, Newton
and Kinnersley), then traces the theory of charged air
particles used to explain the phenomenon from Cavallo
(1777) through the nineteenth century (including
Faraday and Maxwell), the opposing theory that the
effect was due to charged dust particles, and the overthrow of this with the discovery of gas ions in 1896.

697. Roller, Duane, and Duane H.D. Roller, *The Development
of the Concept of Electric Charge: Electricity from
the Greeks to Coulomb.* (Harvard Case Studies in Experimental Science, Case 8.) Cambridge, Mass.:
Harvard University Press, 1954. vi + 97 pp.

Follows a roughly chronological order with liberal
excerpts from primary sources. Concerned mainly with
eighteenth-century work, the larger part of the discussion being devoted to Hauksbee, Gray, Dufay and
Franklin. In common with most works of this period,
the flourishing continental schools--most notably that
of Nollet--are largely ignored. Discusses at some
length the relative merits of the one-fluid and two-fluid
theories.

698. Schofield, Robert E., "Electrical Researches of Joseph
Priestley." *Arch. Int. Hist. Sci.* 16 (1963): 277-
286.

Describes the qualitative researches carried out by
Priestley between 1765 and 1770 and recounted in his
History of Electricity. Though they influenced William
Herschel and probably Henry Cavendish, they were very
soon forgotten.

699. Schonland, Basil F.J., *The Flight of Thunderbolts.* 2nd
ed., revised and enlarged. Oxford: Clarendon Press,
1964. 182 pp.

Contains an interesting description of the first
development of the lightning rod by Franklin and others,
the theories conjectured to explain its workings, and
the controversies that followed as to whether it was

efficacious or dangerous. The bulk of the work describes
the physical processes in lightning.

699a. Smolka, Josef, "B. Franklin, P. Diviš et la découverte
du paratonnerre." *Proceedings, 10th International
Congress of the History of Science, Ithaca, 1962*
(Paris, 1964), Vol. 2, pp. 763-767.

A new assessment of Diviš' work, based on his correspondence with Oetinger and Euler. Concludes that Diviš
was certainly influenced by Franklin's ideas, though
probably only at secondhand, but that some of the conceptions underlying his erection of his lightning rod
were certainly his own.

*   Snelders, H.A.M., "Physics and Chemistry in the Netherlands in the Period 1750-1850."

Includes a discussion of the work of J.H. Van Swinden
(1746-1823) on electricity and magnetism. Cited herein
as item 263.

*   Snorrason, E., *C.G. Kratzenstein ... and His Studies
on Electricity during the Eighteenth Century.*

Cited herein as item 88.

700.   Sudduth, William M., "Eighteenth-Century Identifications
of Electricity with Phlogiston." *Ambix* 25 (1978):
131-147.

Describes various identifications in the period *c.*
1760-*c.* 1790 of electricity with fire and phlogiston,
this generally being made on a basis of comparison of
supposedly similar (chemical) properties. Briefly
discusses the ideas of (among others) Beccaria, Priestley and Van Marum.

701.   Takahashi, Junji, "Change of the Concept of Electric
Current from Galvani to Volta." *Proceedings, 14th
International Congress of the History of Science,
Tokyo, 1974* (Tokyo, 1975), Vol. 2, pp. 327-330.

Contrasts Galvani's idea of static electric charge
stored in an animal with Volta's idea of a continuous
flow which led him to postulate that the source of the
flow was in the conductor.

702.   Thornton, John L., "Charles Hunnings Wilkinson (1763
or 64-1850)." *Ann. Sci.* 23 (1967): 277-286.

A biography and bibliography of a lecturer on electricity who was also a pioneer in medical electricity. Includes brief descriptions of the lectures Wilkinson offered *c.* 1800.

*   Torlais, Jean, *Un physicien au siècle des lumières: l'abbé Nollet, 1700-1770.*

Cited herein as item 96.

703.   Torlais, Jean, "Une grande controverse scientifique au XVIIIe siècle: l'abbé Nollet et Benjamin Franklin." *Rev. Hist. Sci.* 9 (1956): 339-349.

Describes the controversy that divided the French electricians following the translation by opponents of Nollet of Franklin's *Letters on Electricity* in 1752.

704.   Torlais, Jean, "Qui a inventé la bouteille de Leyde?" *Rev. Hist. Sci.* 16 (1963): 211-219.

After examining the claims of various individuals, adopts an agnostic position.

705.   Tunbridge, Paul A., "Franklin's Pointed Lightning Conductor." *Notes Rec. Roy. Soc. Lond.* 28 (1974): 207-219.

Reprints a letter written by Franklin to Horace Benedict de Saussure, sending him an early version of his "separate report" to the 1772 Royal Society committee investigating the relative merits of pointed and blunt conductors.

706.   Walker, W. Cameron, "The Detection and Estimation of Electric Charges in the Eighteenth Century." *Ann. Sci.* 1 (1936): 66-100.

Describes some of the practical and theoretical developments in relation to the electrometer and its derivatives, the condenser and doubler, especially during the latter part of the century. Discusses the work of, *inter alia*, Hauksbee, Gray, Canton, Cavallo, Cavendish, Bennet and Volta. Suggests that Volta's skills in this field played a considerable part in orienting his electrical researches.

707.   Walker, W. Cameron, "Animal Electricity before Galvani." *Ann. Sci.* 2 (1937): 84-113.

Discusses pre-Galvanic work on animal electricity, especially that on the torpedo and similar fish, during

which John Walsh (1772-76) and Henry Cavendish (1776)
established the identity of this animal electricity
with common electricity. Argues that this and Galvani's
own work may have provided an analogy for Volta's construction of his electric pile.

708. Yamazaki, Eizo, "L'abbé Nollet et Benjamin Franklin.
      Une phase finale de la physique cartésienne: la
      théorie de la conservation de l'électricité et de
      l'expérience de Leyde." *Japanese Studs. Hist. Sci.*
      15 (1976): 37-64.

      A sympathetic study of Nollet's writings on electricity
      that emphasizes his naively empiricist attitude towards
      theory construction.

709. Yamazaki, Eizo, "La conduction de l'électricité et
      l'induction électrostatique avant Franklin." *Japanese
      Studs. Hist. Sci.* 16 (1977): 43-49.

      A brief account of the problem posed by induction
      effects in early eighteenth-century investigations of
      electricity.

VI. PHYSICS IN TRANSITION, 1790-1820

(a) Developments in France

(i) General

710. Andoyer, Henri, *L'oeuvre scientifique de Laplace*. Paris: Payot, 1922. 162 pp.

   A brief "modernized précis" of Laplace's chief works.

711. Arnold, David H., "The *Méchanique Physique* of Siméon Denis Poisson: The Evolution and Isolation in France of his Approach to Physical Theory (1800-40)." Ph.D. dissertation, University of Toronto, 1978.

   Discusses the almost total commitment of Poisson to investigating how mathematics could best be applied to physics, arguing that his adherence to the Laplacian ideal for physics and his opposition to the anti-Laplacians can be attributed to doubts on his part as to whether contemporary analysis could adequately represent physical phenomena, especially when it came to taking the discrete structure of matter into account.

712. Bellone, Enrico, "Il significato metodologico dell' eliminazione dei modelli del calorico promossa da Joseph Fourier." *Physis* 9 (1967): 301-310.

   Shows the methodological interconnections that Fourier established in his *Théorie analytique de la chaleur* between his radical viewpoint on the various models of caloric then current and his thesis that a deductive system concerning the "effects of heat" could not be reduced to mechanics.

*   Brown, Theodore, "The Electric Current in Early Nineteenth-Century French Physics."

   Cited herein as item 1082.

*   Bucciarelli, Louis L., and Nancy Dworsky, *Sophie Germain:*
        An Essay in the Theory of Elasticity.

        Cited herein as item 916.

713.   [Bugge, Thomas]. *Science in France in the Revolutionary*
        Era, described by Thomas Bugge, Danish Astronomer Royal
        and Member of the International Commission on the
        Metric System (1798-1799), ed. with introduction and
        commentary by Maurice P. Crosland. Cambridge, Mass./
        London: Society for the History of Technology and
        M.I.T. Press, 1969. xvi + 239 pp.

        Gives a useful overview of French scientific institu-
        tions *ca*. 1798-1799. Includes valuable comments on,
        inter alia, the Ecole Polytechnique, lectures on ex-
        perimental physics by J.A.C. Charles, and the French
        instrument makers of the period.

714.   Cawood, John A., "The Scientific Work of D.F.J. Arago,
        1786-1853." Ph.D. dissertation, University of Leeds,
        1974. xv + 346 pp.

        While a brief biography is provided, the bulk of this
        useful study is devoted to Arago's optical and electro-
        magnetic work including a discussion of his role in the
        introduction of Fresnel's wave theory of light. Also
        deals with his work in the "environmental sciences" and
        has a chapter on "Arago's Approach to Science."

715.   Chaldecott, J.A., "Scientific Activities in Paris in
        1791: Evidence from the Diaries of Sir James Hall
        for 1791, and Other Contemporary Records." *Ann. Sci.*
        24 (1968): 21-52.

        Draws accounts of some electrical experiments of
        Charles' and of a heat regulating device from Hall's
        diary, and in addition discusses the work of the Com-
        mission of Weights and Measures.

716.   Chappert, A., "Deux lettres autographes d'Augustin
        Fresnel." *Arch. Int. Hist. Sci.* 26 (1976): 268-279.

        The first letter (26 Oct. 1815) is to Arago discussing
        a paper Fresnel had submitted to the Academy of Sciences on
        his theory of light; the second (18 Sept. 1821) is con-
        cerned with the design of lenses for lighthouses. Both
        have commentary and notes by Chappert.

717.   Chappert, André, *Etienne Louis Malus (1775-1812) et la*
        théorie corpusculaire de la lumière: traitement analy-

*tique de l'optique géométrique, polarisation de la
lumière et tentative d'explication dynamique de la
réflexion et de la réfraction.* Paris: J. Vrin, 1977.
283 pp.

Gives a brief biography and a condensed history of
18th-century optics before analyzing Malus' optical
research in detail. Concentrates on the theoretical
aspects of his work. Finishes with a brief 'look at
Malus' philosophy of science. Reviewed by A.E. Shapiro,
Isis 69 (1978): 466-467, and R.H. Silliman, *Ann. Sci.* 35
(1978): 434-436.

718. Chappert, A., "Lettres nouvelles de la correspondance
de Fresnel." *Arch. Int. Hist. Sci.* 28 (1978): 49-65.

Provides the background for and the text of a letter
from Fresnel to Biot and Biot's reply concerning the
former's "Mémoire sur la diffraction de la lumière,"
and also a letter from Fresnel to Arago concerning,
amongst other things, magnetism.

719. Chappert, A., "A propos de 'L'essai autobiographique'
de Thomas Young." *Arch. Int. Hist. Sci.* 30 (1980):
102-110.

Comments on Young's recently discovered autobiography
(item 767), emphasizing the new details it provides of
Young's contribution towards making the work of early
nineteenth-century French physicists known in England.

720. Costabel, Pierre, "L'activité scientifique d'Ampère."
Rev. Hist. Sci. 30 (1977): 105-112.

A useful overview of Ampère's scientific work including
brief discussions of his mathematics, optics and heat,
demonstrating his powerful blending of mathematical
theory and experiment.

721. Coyac, J., et al., eds., *Gay-Lussac: la carrière et
l'oeuvre d'un chimiste français durant la première
moitié du XIXe siècle.* Palaiseau: Ecole Polytechnique,
1980. v + 290 pp. (Actes du colloque Gay-Lussac,
11-13 décembre 1978).

A collection of 18 of the papers presented at a conference celebrating the bicentenary of Gay-Lussac's
birth. The contributions vary considerably in quality.
Those of most relevance to the history of physics are
as follows: M. Fétizon, "La thermodynamique de Galilée

à Clausius: influence de Gay-Lussac sur Carnot"; P. Lervig, "Les expériences de Gay-Lussac sur l'expansion adiabatique des gaz"; and J. Rosmorduc, "Les idées de Gay-Lussac sur la lumière." J. Mandelbaum, "Gay-Lussac et la Société Philomathique," includes valuable bibliographical information on the Society, a very important institution for the French science of the period. Reviewed by W.A. Smeaton, *Ann. Sci.* 39 (1982): 77-78.

722.   Crosland, Maurice, *The Society of Arcueil: A View of French Science at the Time of Napoleon I.* London: Heinemann, 1967. xx + 514 pp.

An extremely important study of the most influential group within the French scientific community in the Napoleonic period. Emphasizes the role of patronage, wielded in this case by Laplace and Berthollet, in shaping the careers of younger members of the group. Includes much information about both individuals and institutions.

*   Crosland, Maurice, *Gay-Lussac, Scientist and Bourgeois.*

Cited herein as item 45.

723.   Finn, Bernard S., "Laplace and the Speed of Sound." *Isis* 55 (1964): 7-19.

Describes Laplace's early recognition that there must be an error in Newton's theory of the transmission of sound, and his recognition of the relevance of the ratio of the specific heats of air (c_p/c_v) to the calculation. Further describes Laplace's attempts to link the theory to several modifications of his caloric theory of heat during the period 1802-1823.

*   Fox, Robert, *The Caloric Theory of Gases from Lavoisier to Regnault.*

Cited herein as item 333.

724.   Fox, Robert, "The Rise and Fall of Laplacian Physics." *Hist. Studs. Phys. Sci.* 4 (1974): 89-136.

Argues that between 1790 and 1820 there was a physical research program dominating French science, centering on the Arcueil group under Laplace and Berthollet and aiming to reduce all phenomena to central forces between particles whether of matter or of imponderable fluids. This program was subsequently overthrown by the work of

Fresnel, Arago, Fourier, Dulong and Petit in the years 1815-1825, without any unified alternative tradition being established in its place.

725. Frankel, Eugene, "Jean-Baptiste Biot: The Career of a Physicist in Nineteenth Century France." Ph.D. dissertation, Princeton University, 1972.

An account of this important protégé of the Laplacian school, describing his early work in magnetism and electricity and his involvement in the extensive optical research of the time leading to his major work on polarization. Also given extensive coverage is Biot's relationship with Laplace and his colleagues, the introduction of the new undulatory theory of light by Fresnel, and Biot's response to the theory and his subsequent loss of prestige and influence.

726. Frankel, Eugene, "The Search for a Corpuscular Theory of Double Refraction: Malus, Laplace and the Prize Competition of 1808." *Centaurus* 18 (1974): 223-245.

Argues that among the French it was Malus who first noticed Wollaston's experimental confirmation of the law of double refraction which Huygens had derived from the wave theory of light; that the 1808 competition was set to allow Malus to explain this in corpuscular terms; and that Laplace and Malus then mathematically replicated Huygens' law from corpuscularian principles, but only by the introduction of arbitrary constructions.

727. Frankel, Eugene, "Corpuscular Optics and the Wave Theory of Light: The Science and Politics of a Revolution in Physics." *Soc. Studs. Sci.* 6 (1976): 141-184.

Shows that in the decade 1805-1815, the corpuscular theory of light gained considerably in strength, especially through the work of Malus and Biot. Particular attention is paid to Biot's extension of the corpuscular theory of polarization to account for coloration effects. Describes the corpuscular response to the wave theory, stressing the importance of Arago's publicizing efforts on behalf of the latter. Argues that the controversy was ultimately resolved "not through the conversion of the corpuscularians, but rather through the ascendancy of an anti-Laplace faction to dominant positions in French science."

728.   Frankel, Eugene, "J.B. Biot and the Mathematization of
       Experimental Physics in Napoleonic France." *Hist.
       Studs. Phys. Sci.* 8 (1977): 33-72.

       Argues that there was a self-conscious attempt around
       1800-1810, especially among the younger, mathematically
       trained members of the Arcueil circle, to find mathe-
       matical laws for the various qualitative fields of
       physics. This is exemplified by both the diversity and
       the intent of Biot's researches in this period into
       electricity, optics, acoustics and magnetism.

729.   Frankel, Eugene, "Career-Making in Post-Revolutionary
       France: The Case of Jean-Baptiste Biot." *Brit. J.
       Hist. Sci.* 11 (1978): 36-48.

       Due to the influence of members of the Institut de
       France over nominations for the new scientific positions
       which opened up in France during this period, winning
       patronage from senior members of the Institut was
       vitally important for advancement. Biot's career,
       traced here, demonstrates that it was self-conscious
       career-making and well-directed hard work rather than
       research excellence that gave him success.

730.   Friedman, Robert Marc, "The Creation of a New Science:
       Joseph Fourier's Analytical Theory of Heat." *Hist.
       Studs. Phys. Sci.* 8 (1977): 73-99.

       Discusses how Fourier, by restricting himself to
       the analysis of heat propagation and avoiding general
       theories of heat, managed to develop his theory as an
       autonomous branch of mathematical physics. Further
       describes the conceptual, physical, methodological and
       epistemological bases of Fourier's theory.

731.   Gillispie, Charles Coulston, "The *Encyclopédie* and the
       Jacobin Philosophy of Science: A Study in Ideas and
       Consequences." *Critical Problems in the History of
       Science*, ed. Marshall Clagett. Madison: University
       of Wisconsin Press, 1959, pp. 255-289.

       Suggests that the Jacobin attack on the Academy of
       Sciences and on the new chemistry was symptomatic of
       the hostility born of a fundamental incompatibility be-
       tween their ideas drawn from Rousseau and Diderot of an
       "organic" nature, and the mathematical Newtonian ideal.
       The argument is disputed by L. Pearce Williams in the
       same volume (item 750).

732. Gillispie, Charles C., and A.P. Yushkevich, *Lazare Carnot, Savant*. Princeton, N.J.: Princeton University Press, 1971. xii + 359 pp.

Contains a brief biography of Carnot and an analysis and subsequent history of his generally ignored work on mechanics, especially his development of ideas of work and power. Shows the inter-relation of his mathematical ideas with his awareness of the constraints of physical reality, and argues that this influenced Carnot's son Sadi, and others such as Coriolis, Cauchy and Poncelet. Yushkevich's section is devoted to Carnot's early work on limit theory.

733. Gillmor, C. Stewart, *Coulomb and the Evolution of Physics and Engineering in Eighteenth-Century France*. Princeton, N.J.: Princeton University Press, 1971. xviii + 328 pp.

A well-researched biography, together with analyses of Coulomb's various contributions to physics and engineering. Argues that mathematical physics emerged in late 18th-century France from an amalgamation of three previously separate traditions, namely: (1) rational mechanics, (2) experimental physics, and (3) practical engineering; and that Coulomb, with his strong background in engineering, played a seminal role in this.

734. Grattan-Guinness, I., "Joseph Fourier and the Revolution in Mathematical Physics." *J. Inst. Maths. Applics.* 5 (1969): 230-253.

A discussion of the major features of Fourier's first (1807) paper on heat conduction, subsequently published in full by the author in collaboration with J.R. Ravetz (item 735).

735. Grattan-Guinness, Ivor, in collaboration with J.R. Ravetz, *Joseph Fourier 1768-1830: A Survey of His Life and Work, based on a Critical Edition of His Monograph on the Propagation of Heat, Presented to the Institut de France in 1807*. Cambridge, Mass./London: M.I.T. Press, 1972. xii + 516 pp.

The first publication of Fourier's first major work, with an accompanying commentary in English. The commentary includes synopses and analyses of particular points of interest in the work and comparisons with his later writings. It is supplemented by a brief biography, appraisals of Fourier's related mathematical work, and

accounts of similar developments by his contemporaries
and Fourier's conflicts with them. Reviewed by T.L.
Hankins, *Isis* 64 (1973): 424-425.

736. Grattan-Guinness, I., "On Joseph Fourier: The Man, the
       Mathematician, and the Physicist." *Ann. Sci.* 32
       (1975): 503-514.

       An essay review of John Herivel's biography of Fourier
       (item 62) which criticizes his interpretation of Fourier's
       physics on the grounds of an insufficient appreciation
       of both Fourier's theory of the calculus and the role
       it played in his work.

737. Grattan-Guinness, I., "Recent Researches in French
       Mathematical Physics of the Early 19th Century."
       Ann. Sci. 38 (1981): 663-690.

       An essay review of six recent publications on early
       19th-century mathematics and/or physics, including in-
       teresting additional information on personal relations
       between the French mathematicians of the period together
       with the texts of several important new documents.

738. Grattan-Guinness, I., "Mathematical Physics in France,
       1800-1840: Knowledge, Activity, and Historiography."
       *Mathematical Perspectives: Essays on Mathematics and
       Its Historical Development*, ed. Joseph W. Dauben.
       New York: Academic Press, 1981, pp. 95-138.

       A major survey of the development of mathematical
       physics during a crucial period in its history. Sug-
       gests a parallel between the evolution of this subject
       from a science based wholly on Newtonian principles of
       particulate interactions into a much more broadly based
       science in which apparently non-Newtonian phenomena
       were also subjected to mathematical treatment, and the
       simultaneous evolution of mathematics itself in which
       18th-century algebraic calculus was superseded by
       mathematical analysis, unified under the theory of
       limits. Also includes valuable information on institu-
       tional aspects of French science and brief remarks on
       the status of mathematical physics in other countries
       at this time. Good bibliography.

739. Grattan-Guinness, I., "Mathematical Physics in France,
       1800-35." *Epistemological and Social Problems of the
       Sciences in the Early Nineteenth Century*, ed. H.N.
       Jahnke and M. Otte. Dordrecht: D. Reidel, 1981, pp.
       349-370.

Pursues similar themes to those developed in item 738.

* Herivel, John, *Joseph Fourier: The Man and the Physicist*.

Cited herein as item 62.

740. Herivel, John, *Joseph Fourier face aux objections contre sa théorie de la chaleur: lettres inédites 1808-1816*. Paris: Bibliothèque Nationale, 1980. 87 pp.

Publishes various manuscripts of Fourier's, mostly concerning his first, 1807, paper on heat diffusion. Also includes extracts from writings of Laplace, Biot and Poisson. Not always reliable English translations of the Fourier documents are given in Herivel's biography of Fourier (item 62).

741. Marcovich, André, "La théorie philosophique des rapports d'André-Marie Ampère." *Rev. Hist. Sci.* 30 (1977): 119-123.

A brief description of one aspect of Ampère's theory of knowledge.

742. Métivier, Michel, Pierre Costabel and Pierre Dugac, eds., *Siméon-Denis Poisson et la science de son temps*. Paris: Ecole Polytechnique, 1981. 285 pp.

A volume commemorating the bicentenary of Poisson's birth. Contains the following articles: (1) Pierre Costabel, "Siméon-Denis Poisson, aspect de l'homme et de son oeuvre" (the French original of Costabel's entry on Poisson in the *Dictionary of Scientific Biography*); (2) David H. Arnold, "Poisson and Mechanics"; (3) Paul Brouzeng, "Poisson et la capillarité selon Duhem d'après un manuscrit inédit: Les leçons sur les théories de la capillarité"; (4) B. Bru, "Poisson, le calcul des probabilités et l'instruction publique"; (5) Louis L. Bucciarelli, "Poisson and the mechanics of elastic surfaces"; (6) André Chappert, "Poisson et les problèmes de l'optique: la controverse avec Fresnel"; (7) E. Coumet, "S.D. Poisson élève à l'Ecole polytechnique: quelques documents inédits"; (8) S.S. Demidov, "Des parenthèses de Poisson aux algèbres de Lie"; (9) R.W. Home, "Physical principles and the possibility of a mathematical science of electricity and magnetism"; (10) Robin E. Rider, "Poisson and algebra: against an 18th-century background"; (11) O.B. Sheynin, "Poisson and statistics"; (12) A.P. Yushkevich, "S.D. Poisson et la théorie de

l'intégration." Also includes an annotated list of
Poisson's publications, sorted into a much more useful
arrangement than in previously published lists.

743. Pizzamiglio, Pier Luigi, "La nascita della fisica: in
      merito ad una proposta di W.F. Cannon." *Physis* 18
      (1976): 165-173.

      Adopting Cannon's thesis (in item 789) that physics
      as a discipline first evolved, in France, c. 1810-30,
      examines the development of the text books of E.G.
      Fischer, R.J. Haüy and J.B. Biot. The contrast between
      Biot's texts and those of Haüy is stressed.

744. Ravetz, J.R., "Preliminary Notes on the Study of J.B.J.
      Fourier." *Arch. Int. Hist. Sci.* 13 (1960): 247-251.

      Brief overview of Fourier's fundamental innovations,
      from his 1807 paper onward, in the method and application
      of mathematical physics and on the nature of dimensions
      and physical constants.

745. Rosmorduc, Jean, "Ampère et l'optique: une intervention
      dans le débat sur la transversalité de la vibration
      lumineuse." *Rev. Hist. Sci.* 30 (1977): 159-167.

      Provides a brief account of the acceptance of Fres-
      nel's theory, and shows the importance of both Ampère's
      support and his co-operation in developing the theory
      of transverse waves.

746. Sadoun-Goupil, Michelle, "Esquisse de l'oeuvre d'Ampère
      en chimie." *Rev. Hist. Sci.* 30 (1977): 125-141.

      Includes a description of Ampère's theory of molecular
      structures, derived from his electrochemical work.

747. Silliman, Robert H., "Fresnel and the Emergence of
      Physics as a Discipline." *Hist. Studs. Phys. Sci.* 4
      (1974): 137-162.

      Argues that Fresnel's work in optics is an exemplar
      of a new, sophisticated methodology of physics which
      united the separate older traditions of *physique par-
      ticulière* and *physique générale*, and that this new com-
      bination of experiment and mathematics led to the first
      successful attack on imponderable fluids and turned
      attention away from matter to motion in the construction
      of physical models.

748. Speziali, Pierre, "Augustin Fresnel et les savants genevois." *Rev. Hist. Sci.* 10 (1957): 255-259.

    A previously unpublished letter of Fresnel to M.A. Pictet of February, 1823, describing aspects of his past work.

749. Sutton, Geoffrey, "The Politics of Science in Early Napoleonic France: The Case of the Voltaic Pile." *Hist. Studs. Phys. Sci.* 11 (1981): 329-366.

    Analyzes the French reaction to Volta's pile, emphasizing the previously neglected work done during the year preceding Volta's visit to Paris in the fall of 1801. Argues that initially there were two competing approaches, that of analytical chemists such as Fourcroy and Guyton de Morveau on the one hand, and that of Laplace and his associates on the other, the latter of which tried to reduce the operation of the pile to Coulomb-style electrostatic forces. Napoleon's preferential treatment of Laplace and Berthollet had the incidental effect of institutionalizing the reductionist Laplacian program at the expense of the chemists.

750. Williams, L. Pearce, "The Politics of Science in the French Revolution." *Critical Problems in the History of Science*, ed. Marshall Clagett. Madison: University of Wisconsin Press, 1959, pp. 291-308.

    Argues *contra* Gillispie (item 731) that support for, or opposition to, science cannot be attributed to any particular political party, and that the lack of any successful organizational reform was due to conflicts between rival reforming programmes rather than opposition to science itself.

(ii) Institutional Histories

751. Bradley, Margaret, "The Facilities for Practical Instruction in Science During the Early Years of the Ecole Polytechnique." *Ann. Sci.* 33 (1976): 425-446.

    A review of available evidence to determine how much equipment was available, how it was obtained and how extensively it was used by the students, 1794-*c.* 1812.

752. Brunet, Joseph, Jr., "Science and the Early Ecole Polytechnique, 1794-1806: The Impact of the Early *Polytechniciens* on the Science of the 18th Century and on

the Industrial Revolution in the 19th Century." Ph.D. dissertation, University of Kentucky, 1969.

Discusses the formation and early development of the Ecole Polytechnique, describes some of the courses taught there and follows the achievements of some of its more prominent graduates.

*   [Bugge, Thomas]. "Science in France in the Revolutionary Era."

      Cited herein as item 713.

753.   Crosland, M.P., "The Development of a Professional Career in Science in France." *The Emergence of Science in Western Europe*, ed. M.P. Crosland. London: Macmillan, 1975, pp. 139-160. Also published in *Minerva* 13 (1975): 38-57.

      Concentrates on Revolutionary and post-Revolutionary Paris, showing the increasing opportunities for paid scientific research and training, and the increasingly structured nature of the institutions of French science.

754.   Hahn, Roger, "The Problems of the French Scientific Community, 1793-1795." *Proceedings, 12th International Congress of the History of Science, Paris, 1968* (Paris, 1971), vol. 3B, pp. 37-40.

      Argues that in 1793-1795 there briefly existed a "democratic" phase in Parisian science, according to which the advancement of science would be carried on by voluntary associations. These associations soon failed.

755.   Mandelbaum, Jonathan, "La Société philomathique de Paris: recherches sur le personnel et les publications d'une société scientifique parisienne de 1788 à 1814." Mémoire pour le Diplome, Ecole des Hautes Etudes en Sciences Sociales, Paris, 1976.

      Sheds much light on an important but little known institution.

*   Mandelbaum, Jonathan, "Gay-Lussac et la Société Philo-mathique."

      Cited herein in item 721.

756.   Outram, Dorinda, "Politics and Vocation: French Science, 1793-1830." *Brit. J. Hist. Sci.* 13 (1980): 27-43.

Argues that standard models of professionalization and institutionalization are of limited use in assessing the rise of science in the period in question, due to the non-autonomous nature of the discipline and the political involvement of many of the patrons of science and also their opponents.

757. Williams, L. Pearce, "Science, Education and the French Revolution." *Isis* 44 (1953): 311-330.

Argues that after the anarchic state of educational policy in the early stages of the Revolution, the sciences were included in the curriculum of the Ecoles centrales, 1795-1799, to train students in precise habits of thought and against superstition as part of a course in Condillac's "Ideology." However, science subjects proved popular only if demonstration apparatus were available or if they displayed practical utility. The course as a whole failed.

(b) Developments Elsewhere

758. Caneva, Kenneth L., "From Galvanism to Electrodynamics: The Transformation of German Physics and Its Social Context." *Hist. Studs. Phys. Sci.* 9 (1978): 63-159.

On the basis of German writings on electricity and magnetism from the first half of the 19th century, delineates two very different styles in German physics during this period, namely "concretizing science," qualitative in character and asserting experience as the direct source of theory, and "abstracting science," quantitative, mathematical and hypothetico-deductive in structure. Argues that these were maintained by two different generational groups, and links the change in scientific outlook from one generation to the next with wider changes in German culture and society following the Napoleonic invasions.

759. Cantor, G.N., "Thomas Young's Lectures at the Royal Institution." *Notes Rec. Roy. Soc. Lond.* 25 (1970): 87-112.

Describes Young's appointment as lecturer at the Royal Institution (1801), his lack of success especially compared with the engaging Davy, and his resignation (1803).

Then, drawing on the notebooks of Young's lecture courses, discusses his natural philosophy, especially his rapidly changing, *ad hoc* aether theories which he used to explain cohesion and gravitation and as a basis for his optical theory.

760. Cantor, G.N., "The Changing Role of Young's Ether." *Brit. J. Hist. Sci.* 5 (1970-71): 44-62.

Shows how Young, between 1799 and 1807, first aimed to account for all natural phenomena by means of a number of subtle fluids freely transmissible through matter but also attracted to matter particles, and later rejected most of these ideas, eventually retaining only the luminiferous aether.

761. Cantor, G.N., "Henry Brougham and the Scottish Methodological Tradition." *Studs. Hist. Phil. Sci.* 2 (1971): 69-89.

Suggests that Brougham's rejection of Young's wave theory of light was due to his belief in the Scottish methodological tradition "which eschewed conjectures concerning unobservable causes," including aetherial fluids. Argues that this tradition derived largely from Reid's strongly inductivist methodology.

762. Cantor, G.N., "The Reception of the Wave Theory of Light in Britain: A Case Study Illustrating the Role of Methodology in Scientific Debate." *Hist. Studs. Phys. Sci.* 6 (1975): 109-132.

Argues that the different British responses to the wave theory indicates the distinction between the older "Newtonian inductivist" common-sense school which emphasized experiment and a unified natural philosophy, and the new Cambridge-based, mathematically oriented methodology that stressed bold hypotheses and mathematical and mechanical world views.

763. Cook, Thomas Hyde, "Science, Philosophy and Culture in the Early *Edinburgh Review*, 1802-29." Ph.D. dissertation, University of Edinburgh, 1976.

Uses the *Review* as an "indication of developmental patterns in intellectual and cultural history." Ties the contributors to the *Review* into the general philosophical and cultural background of contemporary Edinburgh. Includes a section on the physical science contributions of Brougham, Playfair and Leslie.

764. Crosland, Maurice, and Crosbie Smith, "The Transmission
      of Physics from France to Britain: 1800–1840." *Hist.
      Studs. Phys. Sci.* 9 (1978): 1–61.

      Shows how, after an initially cautious British recep-
      tion of early 19th-century French physics, the mathe-
      matical approach of Biot, Fourier, Poisson, and others
      was incorporated into the Cambridge mathematical tradi-
      tion where it further stimulated three different "pro-
      gressive conceptual bases," namely Laplacian reduction
      to attractive and repulsive forces; the resolution of
      phenomena into aetherial and material components; and
      an emphasis on pure mathematical law applied to observable
      rather than inferred underlying entities.

765. Gower, Barry S., "Speculation in Physics: The History
      and Practice of *Naturphilosophie*." *Studs. Hist. Phil.
      Sci.* 3 (1973): 301–356.

      Following a discussion of the ideas of Kant, Fichte
      and especially Schelling, a general *Naturphilosoph*
      metaphysic is attributed to Ritter and Oersted, charac-
      terized in their scientific speculations by their con-
      ception of the essential unity of chemical, electrical
      and magnetic actions, these being held to be manifesta-
      tions of different combinations and transformations of
      the same basic "forces" differently "polarized." Stresses
      that this provides merely a conceptual framework and is
      not explanatory in itself.

766. Heidelberger, Michael, "Some Patterns of Change in the
      Baconian Sciences of Early 19th Century Germany."
      *Epistemological and Social Problems of the Sciences
      in the Early Nineteenth Century*, ed. H.N. Jahnke and
      M. Otte. Dordrecht: D. Reidel, 1981, pp. 3–18.

      Using Ohm's work on electricity as an example, argues
      that a major transformation took place in the general
      conception of science in Germany during the first half
      of the 19th century. Sees Ohm's work as closely paral-
      leling Fourier's phenomenological treatment of heat
      conduction.

767. Hilts, Victor L., "Thomas Young's Autobiographical
      Sketch." *Proc. Amer. Phil. Soc.* 122 (1978): 248–260.

      Publishes the text of brief newly rediscovered auto-
      biographical notes of Young's.

768. Knight, David M., "The Physical Sciences and the Roman-
      tic Movement." *Hist. Sci.* 9 (1970): 54-75.

      Stresses that it is necessary to be cautious about
      assuming a causal interaction between two intellectual
      fields--here Romanticism and physics--where there may
      in reality be only a parallelism in vocabulary and
      methodology derived from a similar intellectual heritage.

769. Knight, David M., "German Science in the Romantic
      Period." *The Emergence of Science in Western Europe*,
      ed. M.P. Crosland. London: Macmillan, 1973, pp. 161-
      178.

      Discusses the distinct characteristics of German
      science *c.* 1780-1839, arguing for both a *Naturphilosophie*
      tradition and an equally strong empiricist and generalist
      tradition. A useful discussion of recent secondary
      literature.

769a. Latchford, K.A., "Thomas Young's Work on Optics." Ph.D.
      thesis, University of London, 1975.

      Not seen.

770. Lilley, S., "Nicholson's Journal, 1797-1813." *Ann. Sci.*
      6 (1948): 78-101.

      Describes how public interest in science and the need
      for alternative publication sources created room for
      such a journal, and how, after publishing the first
      English work on the Voltaic cell, the journal came to
      monopolize publications in this field during the years
      1801-1813.

771. Morrell, J.B., "Professors Robison and Playfair, and
      the *Theophobia Gallica*: Natural Philosophy, Religion
      and Politics in Edinburgh 1789-1815." *Notes Rec.*
      Roy. Soc. Lond. 26 (1971): 43-63.

      The xenophobic anti-Jacobinism of Robison, professor
      of natural philosophy at Edinburgh, who condemned
      Lavoisierian chemistry, the metric system, Laplacian
      astronomy and Priestley's aether theories on the grounds
      of their atheism and connections with Jacobinism, is
      contrasted with the approach of his Whig successor,
      Playfair, who openly supported French mathematical and
      physical science, rejecting all forms of dogmatism.

772. Morse, Edgar William, "Natural Philosophy, Hypotheses, and Impiety: Sir David Brewster Confronts the Undulatory Theory of Light." Ph.D. dissertation, University of California, Berkeley, 1972.

Describes the reasons why Brewster maintained his allegiance to the Newtonian theory of light, namely his view that the undulatory theory was founded on a hypothetical entity, his general rejection of speculative science as counter to God's will, his lack of sympathy with mathematical theories, and the fact that, initially at least, the undulatory theory had problems in explaining some phenomena.

772a. Schimank, Hans, "Ludwig Wilhelm Gilbert und die Anfänge der *Annalen der Physik.*" *Sudhoffs Archiv* 47 (1963): 360-372.

Provides valuable details concerning Gilbert's life and work, especially in connection with his editing of the *Annalen*, 1799-1824.

773. Siegfried, Robert, "Boscovich and Davy: Some Cautionary Remarks." *Isis* 58 (1967): 236-238.

Argues, against L. Pearce Williams' interpretation, that there is no clear evidence of Davy having viewed Boscovich's point atomism as anything more than one of many possibly helpful ideas.

774. Snelders, H.A.M., "The Influence of the Dualistic System of Jakob Joseph Winterl (1732-1809) on the German Romantic Era." *Isis* 61 (1970): 231-240.

Describes Winterl's dualist conception of matter, its relationship with the general temper of German Romanticism and *Naturphilosophie*, its influence, and its downfall due to its lack of connection with empirical evidence.

775. Snelders, H.A.M., "J.S.C. Schweigger: His Romanticism and His Crystal Electrical Theory of Matter." *Isis* 62 (1971): 328-338.

Describes Schweigger's theory (first conceived *c.* 1808) that "atoms" of elements had positive and negative poles and formed electrically neutral crystals. There were many problems with the theory, however, and it gained no following.

776. Van Broeckhoven, R.L.J.M., "The Growth of Thomas Young's Ideas on Interference." *Proceedings, 13th International*

Congress of the History of Science, Moscow, 1971 (Moscow, 1974), vol. 6, pp. 324–329.

Traces Young's ideas from his realization that both superposition and path differences of pulses were significant in acoustic resonance.

777. Wetzels, Walter D., *Johann Wilhelm Ritter: Physik im Wirkungsfeld der deutschen Romantik*. Berlin/New York: Walter de Gruyter, 1973. xii + 139 pp.

A careful analysis of the conceptual framework within which Ritter carried out his scientific investigations.

778. Williams, L. Pearce, "Kant, *Naturphilosophie* and Scientific Method." *Foundations of Scientific Method: The Nineteenth Century*, ed. by R.N. Giere and R.S. Westfall. Bloomington/London: Indiana University Press, 1973, pp. 3–22.

Argues for Kant's influence on the method of the *Naturphilosophen* (especially Oersted) by providing justification for the use of speculation, yielding a teleological view of Nature and a physics based on interconvertible forces.

*   Wood, Alexander, *Thomas Young, Natural Philosopher, 1773–1829*.

Cited herein as item 107.

779. Worrall, John, "Thomas Young and the 'Refutation' of Newtonian Optics: A Case-Study of the Interaction of Philosophy of Science and History of Science." *Method and Appraisal in the Physical Sciences: The Critical Background to Modern Science, 1800–1905*, ed. Colin Howson. Cambridge: Cambridge University Press, 1976, pp. 107–179.

Argues convincingly that Young's work was ignored because it did not establish the superiority of the wave theory of light, not because of excessive worship of Newton's corpuscular theory.

VII. NINETEENTH-CENTURY PHYSICS

(a) General

780. Agassi, Joseph, "Sir John Herschel's Philosophy of Success." *Hist. Studs. Phys. Sci.* 1 (1969): 1-36.

"An extended critical book-report" on Herschel's popular, influential and "strictly Baconian" *Preliminary Discourse on the Study of Natural Philosophy* (1831). Briefly argues that this was an attempt to restore order to rapidly changing contemporary science by redescribing what was, in Herschel's opinion, the only foolproof scientific methodology, and by identifying science with scientific success.

781. Badash, Lawrence, "The Completeness of Nineteenth-Century Science." *Isis* 63 (1972): 48-58.

Quoting from addresses of various physicists, Badash argues that there was a generalized but "low-grade" feeling at the end of the century that physical science was approaching completeness. Leaves open the questions of the influence, geographical distribution and context of this "malaise."

782. Basalla, George, William Coleman and Robert Kargon, eds., *Victorian Science: A Self Portrait from the Presidential Addresses of the British Association for the Advancement of Science.* Garden City, N.Y.: Doubleday, Anchor Books/London: G. Bell and Sons, 1970. x + 510 pp.

An edited selection from some seventy addresses providing an outline of Victorian science as seen by some of its leading practitioners. Concerned mostly with the philosophical, methodological, and organizational aspects of science.

783. Becher, Harvey W., "William Whewell and Cambridge Mathe-
      matics." *Hist. Studs. Phys. Sci.* 11 (1980): 1–48.

      Describes the process whereby analysis was introduced
      into the Cambridge mathematics course and, largely due
      to the influence of Whewell, rapidly subordinated to
      applied mathematics. Suggests that the dependence of
      British physicists on geometry and mechanical models
      may owe at least as much to the Cambridge mathematical
      tradition as to the Scottish Common Sense philosophy
      to which Richard Olson (item 823) has drawn attention.

*   Berman, Morris, *Social Change and Scientific Organiza-
      tion: The Royal Institution 1799–1844.*

      Cited herein as item 850.

784. Bork, Alfred M., "The Fourth Dimension in Nineteenth-
      Century Physics." *Isis* 55 (1964): 326–338.

      Describes several scattered speculations by authors
      such as Maxwell, Gibbs, Knott, and especially Clifford
      and C.M. Minton at the end of the 19th century on the
      concept of a four-dimensional mathematical physics.
      Argues that this may indicate a dissatisfaction with
      the traditional mechanical models and a turning to those
      based on abstract mathematical systems.

784a. Brock, W.H., N.D. McMillan and R.C. Mollan, eds., *John
      Tyndall: Essays on a Natural Philosopher.* Dublin: Royal
      Dublin Society, 1981.

      Not seen.

785. Brush, Stephen G., *The Temperature of History: Phases
      of Science and Culture in the Nineteenth Century.*
      New York: Burt Franklin, 1978. (Studies in the
      History of Science, Volume 4.) xii + 210 pp.

      A collection of eight relatively independent essays
      on the interactions between physics (especially the
      theory of heat) and "general culture" in the 19th
      and early 20th centuries. The topics discussed are:
      approaches to the historiography of science; romanticism
      and realism; the dispute over the age of the earth;
      the distinction between fundamental and applied physics,
      and the contrast between the high status of applied
      physics in the 19th century and its low status today;
      the heat death of the universe; testing the efficacy
      of prayer; materialism and the kinetic theory; eugenics
      and the theory of degeneration; and pseudo-thermodynamical
      theories of history. Reviewed by K.R. Hutchison, *Ann.
      Sci.* 37 (1980): 716–718.

786. Cannon, Walter F., "John Herschel and the Idea of
     Science." *J. Hist. Ideas* 22 (1961): 215-239.

     Analyzes Herschel's *Preliminary Discourse* (1830) as
     a definitive statement of contemporary attitudes towards
     science. Concentrates on astronomy, concluding that,
     for Herschel, "to be scientific" meant "to be like
     physical astronomy." Also discusses Herschel's life
     and his attitudes to physics and to methodology in
     general.

787. Cannon, Walter F., "The Normative Role of Science in
     Early Victorian Thought." *J. Hist. Ideas* 25 (1964):
     487-502.

     Argues that science enjoyed a great deal of respect
     in early Victorian society, even among artists and
     theologians, because of the picture of order and secure
     knowledge that it presented.

788. Cannon, Walter F., "History in Depth: The Early Victorian
     Period." *Hist. Sci.* 3 (1964): 20-38.

     Insists on the need to examine the history of science
     as a part of general history, arguing (among other
     things) that the evolution of early Victorian concepts
     and methods into later physical theories must be seen
     in terms of developments in a wider social and intellec-
     tual context.

789. Cannon, Susan F. (formerly Walter F. Cannon), *Science in
     Culture: The Early Victorian Period*. New York: Science
     History Publications/Folkestone: Dawson, 1978. xii +
     296 pp.

     Several chapters are revised versions of previously
     published articles. The aim of the work is to place
     science in the context of other early Victorian ideas
     and attitudes. Topics discussed are: how scientific
     truth was a part of the wider intellectual culture;
     the importance of the "Cambridge Network" (including
     John Herschel, Whewell and Peacock) in establishing
     Victorian attitudes to science; the "Humboldtian" global
     view of phenomena; the origins of the British Associa-
     tion for the Advancement of Science and a critique of
     current historical definitions of professionalization;
     and the fundamental distinction between the "Newtonian
     natural philosophers" (Young, Brewster, Faraday) and
     the new "physicists" (Thomson, Maxwell, Tait and Stokes).

790. Channell, David Francis, "A Unitary Technology: The
     Engineering Science of W.J.M. Rankine." Ph.D. disser-
     tation, Case Western Reserve University, 1975.

From a study of Rankine's influential textbooks, presents him as an engineer who tempered ideal models of
machines and structures with modifications determined
by empirically derived conditions.

791. Cohen, I. Bernard, "Conservation and the Concept of
      Electric Charge: An Aspect of Philosophy in Relation
      to Physics in the Nineteenth Century." *Critical Prob-
      lems in the History of Science*, ed. Marshall Clagett.
      Madison: University of Wisconsin Press, 1959, pp.
      357-383.

      Describes some 19th-century attitudes to the concept
      of conservation, and conservation of charge in particu-
      lar.

792. Cohen, Robert S., and Raymond J. Seeger, eds., *Ernst
      Mach: Physicist and Philosopher*. Dordrecht: D. Reidel,
      1970. (Boston Studies in the Philosophy of Science,
      VI). viii + 295 pp.

      Includes, in addition to biographical and bibliographical
      material, the following historically oriented articles
      on Mach's work as a physicist: Wolfgang F. Merzkirch,
      "Mach's Contribution to the Development of Gas Dynamics";
      Raymond J. Seeger, "On Mach's Curiosity about Shock
      Waves"; Erwin N. Hiebert, "The Genesis of Mach's Early
      Views on Atomism"; and Gerald J. Holton, "Mach, Einstein
      and the Search for Reality."

793. Davie, George Elder, *The Democratic Intellect: Scotland
      and Her Universities in the Nineteenth Century*. Edin-
      burgh: at the University Press, 1961. xx + 372 pp.

      Concerned primarily with general trends in the 19th-
      century Scottish universities--their resistance to
      anglicization, their efforts to cope with modern special-
      izing tendencies, their involvement in the struggle be-
      tween Church and State. Includes, however, a substantial
      section (pp. 105-200) on debates over the style of
      teaching that should be adopted in mathematics and
      physics and, in particular, the extent to which mathematics
      should be taught from a foundational point of view and
      physics from a mathematical one. J.D. Forbes figures
      large in the discussion. The section culminates in a
      new analysis, in the light of what has gone before, of
      Maxwell's philosophical position.

794. Domb, C., "James Clerk Maxwell in London, 1860-1865."
     Notes Rec. Roy. Soc. Lond. 35 (1980): 67-103.

     Gives details of Maxwell's life and work while he was
     professor of natural philosophy at King's College, London,
     including the courses he taught, examinations he set and
     his research on electrical standard units and on the
     viscosity of gases. Corrects previous accounts of the
     reason for his departure from King's.

795. Donnan, F.G., and A. Haas, eds., *A Commentary on the
     Scientific Writings of J. Willard Gibbs*. 2 vols.
     New Haven: Yale University Press, 1936.

     Companion volumes to Gibbs' *Works*, these consist of a
     series of chapters, some highly technical, by different
     authors explaining, surveying, and critically appraising
     various of Gibbs' papers.

796. *Encyklopädie der mathematischen Wissenschaften, mit
     Einschluss ihrer Anwendungen*. 6 vols. in 23. Leipzig:
     B.G. Teubner, 1898-1935.

     While this is not an explicitly historical work, its
     comprehensive nature leads the contributors--among the
     most prominent workers in their fields at the time--to
     discuss many subjects in the light of their historical
     development. Vol. 4, in four parts and edited by
     F. Klein and C. Müller, is on mechanics; vol. 5, in
     three parts and edited by A. Sommerfeld, is on physics.

797. Geymonat, Ludovico, "La classificazione delle scienze
     in Ampère e in Comte." *Physis* 11 (1969): 223-230.

     Compares the roughly contemporary classifications of
     the two men and suggests some reasons for the differences
     between them.

797a. Gossick, B.R., "Heaviside and Kelvin: A Study in Con-
     trasts." *Ann. Sci.* 33 (1976): 275-287.

     Describes the sharply contrasting educational back-
     ground, social circumstances, character and style of
     scientific work of Heaviside and Kelvin. Includes com-
     ments on letters that passed between the two men and on
     other contacts between them.

798. Harman, P.M. (formerly P.M. Heimann), *Energy, Force,
     and Matter: The Conceptual Development of Nineteenth-
     Century Physics*. Cambridge: Cambridge University Press,
     1982. x + 182 pp.

A good clear survey that draws upon a wide range of
the recent historical literature. Concentrates on conceptual issues rather than points of technical detail.

799.   Heimann, P.M., "The *Unseen Universe*: Physics and the
        Philosophy of Nature in Victorian Britain." *Brit. J.
        Hist. Sci.* 6 (1972-73): 73-79.

        Contrasts the general opinion among Victorian physicists
        such as Maxwell and Thomson of a separation between
        the natural world and a non-interventionist God with
        that of Balfour Stewart and P.G. Tait (1875) who advocated
        a Platonic hierarchy of the natural and supernatural in
        the universe, with the lawlike transference of energy
        between the two levels according to conservation of
        energy principles.

800.   Hennemann, Gerhard, *Naturphilosophie im 19. Jahrhundert.*
        Freiburg/Munich: Karl Alber, 1959. 122 pp.

        A series of essays on philosophical responses to
        science and *vice versa*, including arguments to display
        the relationship between *Naturphilosophie* and the work
        of scientists such as Oersted, Mayer and Humboldt. Re-
        viewed by C.C. Gillispie, *Isis* 53 (1962): 273-275.

801.   Herivel, J., "The Influence of Fourier on British Mathe-
        matics." *Centaurus* 17 (1972): 40-57.

        Argues that Fourier's work strongly influenced at
        least William Thomson and through him others such as
        Joule, Maxwell and Heaviside. Thomson was influenced
        not only in his mathematical techniques, but also in
        his stress on the necessary applicability of theory to
        physical reality and especially in his recognition of
        the analogy between temperature and electrical potential
        which ultimately led to Maxwell's electrodynamics.

802.   Herivel, J., "Aspects of French Theoretical Physics in
        the Nineteenth Century." *Brit. J. Hist. Sci.* 3 (1966-67)
        109-132.

        Suggests that the relative non-productivity of French
        physicists was due to (i) their over-involvement in
        politics and administration; (ii) the general anti-
        scientific climate of the nineteenth century; (iii) the
        dominance in France of a positivist outlook at a time
        when elsewhere the successes of nineteenth-century
        physics were being achieved with mechanical models;
        and (iv) the early deaths of Sadi Carnot and Fresnel.

802a.   Hiebert, Erwin N., "The State of Physics at the Turn of
        the Century." *Rutherford and Physics at the Turn of*

the Century, ed. Mario A. Bunge and William R. Shea.
New York: Science History Publications, 1979. pp. 3-22.

Attempts to delineate the state of physics *ca*. 1900 as
it would have appeared to physicists at the time, identifying the major trends, anchor points, and puzzling problems
within the discipline at that period.

803. Jahnke, H.N., and M. Otte, eds., *Epistemological and
Social Problems of the Sciences in the Early Nineteenth
Century*. Dordrecht: D. Reidel, 1981. xlii + 430 pp.

The proceedings of a conference held at the University
of Bielefeld in November 1979. Relevant papers are
noted individually elsewhere (items 739, 766).

804. Jones, R.V., "James Clerk Maxwell at Aberdeen, 1856-
1860." *Notes Rec. Roy. Soc. Lond.* 28 (1973-74):
57-81.

Describes Maxwell's period as professor of natural
philosophy at Marischal College until his dismissal
when the College merged with King's to form the University of Aberdeen.

805. Kargon, Robert H., "William Rowan Hamilton, Michael
Faraday, and the Revival of Boscovichean Atomism."
Amer. J. Phys. 32 (1964): 792-795.

Argues that because of his philosophical idealism,
Hamilton preferred to found his science on the notion
of force rather than of matter and on this basis adopted
Boscovichean atomism, regarding his own work as the
mathematical completion of the model. Argues also
that, as early as 1834, Faraday was similarly interested
in Boscovich.

806. Kargon, Robert H., "Model and Analogy in Victorian
Science: Maxwell's Critique of the French Physicists."
J. Hist. Ideas 30 (1969): 423-436.

Delineates two different schools of thought in nineteenth-century French physics, the "mechanico-molecular"
school (Laplace, Poisson, etc.) who tried to reduce all
of physics to forces acting between particles and, in
opposition, those such as Fourier who sought in positivistic fashion to reduce physics to analysis. Argues
that Maxwell rejected both approaches and sought a *via
media* based on the use of physical analogies, and illustrates his views by reference to his use of mechanical
models in his electrodynamics.

807. Kargon, Robert H., *Science in Victorian Manchester:
Enterprise and Expertise*. Baltimore: Johns Hopkins

University Press, 1977. xi + 283 pp.

Aims to "draw a portrait of the evolution of Manchester science in adapting to its social context and in response to pressure internal to the scientific disciplines" from 1840 to 1910. Uses Manchester as a model of a more general change from the science of the gentleman-amateur to the institutionalized and professionalized discipline of the twentieth century. Does this by examining the scientific, methodological and social writings of the scientists, by looking at the various institutions established, and by investigating the interactions between science, technology and the wider community. Reviewed by G.K. Roberts, *Isis* 68 (1977): 329-330.

808. Kevles, Daniel J., *The Physicists: The History of a Scientific Community in Modern America*. New York: Alfred A. Knopf, 1978. xiv + 489 pp.

Concerned exclusively with the American physics community and then chiefly with its evolution during the twentieth century. The first few chapters do, however, give a useful survey of American physics in the nineteenth century.

*   Klein, Martin J., *Paul Ehrenfest: Vol. 1, The Making of a Theoretical Physicist*.

Cited herein as item 69.

809. Koizumi, Kenkichiro, "The Emergence of Japan's First Physicists, 1868-1900." *Hist. Studs. Phys. Sci.* 6 (1975): 3-108.

Describes the self-conscious rapid absorption of Western science into Japan following the Meiji restoration. The emergence of physics as a separate discipline is traced through both the work of its first practitioners and the development of its institutions, with constant reference to contemporary ideas concerning Westernization in general.

810. Kuhn, Thomas S., *Black-Body Theory and the Quantum Discontinuity, 1894-1912*. Oxford: Clarendon Press/New York: Oxford University Press, 1978. xvi + 356 pp.

Chiefly concerned with tracing the gradual clarification of the quantum idea during the first decade of the 20th century. Also, however, contains a detailed analysis of Planck's intellectual background in late 19th-century thermodynamics, kinetic theory and electromagnetism. Review symposium by Martin J. Klein, Abner Shimony and Trevor J. Pinch, *Isis* 70 (1979): 429-440.

811. Levere, Trevor H., *Affinity and Matter: Elements of Chemical Philosophy 1800-1865*. Oxford: Clarendon Press/New York: Oxford University Press, 1971. xvii + 230 pp.

   Despite the title, the work covers broader territory than chemistry alone. Develops Thackray's work in his *Atoms and Powers* (item 466) by investigating British Newtonianism and natural theology as manifest in matter theory in the first half of the 19th century. Disagrees in particular with L. Pearce Williams' interpretation of Faraday's matter theory (item 105). Reviewed by W.H. Brock, *Isis* 63 (1972): 458-459.

812. Lezhneva, O.A., "Die Entwicklung der Physik in Russland in der ersten Hälfte des 19. Jahrhunderts." *Sowjetische Beiträge zur Geschichte der Naturwissenschaften*, ed. Gerhard Harig. Berlin: VEB Deutschen Verlag der Wissenschaften, 1960, pp. 203-225.

   Describes the work of early nineteenth-century Russian physicists, above all in electricity and geophysics. Focuses especially on the contributions of V.V. Petrov (1761-1834) and E.C. Lenz (1804-1865) to electrical science.

813. McCormmach, Russell, "Editor's Foreword." *Hist. Studs. Phys. Sci.* 3 (1971): ix-xxiv.

   Advocates the *discipline* as a focus for studying the history of a science, and illustrates the value of the perspective thus provided by surveying developments in the discipline of physics in 19th-century Germany.

814. McCormmach, Russell, *Night Thoughts of a Classical Physicist*. Cambridge, Mass.: Harvard University Press, 1982. 219 pp.

   A fascinating evocation of the mental outlook of a classical physicist in a provincial German university as the social and intellectual world he has known collapses at the end of World War I. The approach is fictional, but the account is constructed from opinions expressed by and events that occurred in the lives of real individuals.

815. Maier, Clifford L., *The Role of Spectroscopy in the Acceptance of the Internally Structured Atom, 1860-1920*. New York: Arno Press, 1981.

A 1964 doctoral dissertation from the University of
Wisconsin which is a history of neither atomism nor
spectroscopy, but rather a study of the inter-relations
between spectroscopy and theories of matter in the
period stated.

816.   Marcucci, Silvestro, "Di alcuni contributi di William
        Whewell alla nomenclatura scientifica." *Physis* 5
        (1963): 373-382.

        Looks at Whewell's ideas on the conceptual value of
        developing an adequate scientific vocabulary.

816a.   [Maxwell, James C.], "James Clerk Maxwell's Inaugural
        Lecture at King's College London." *Amer. J. Phys.*
        47 (1979): 928-933.

        The text of Maxwell's 1860 lecture published for the
        first time.

*   Mendelssohn, K., *The World of Walther Nernst: The Rise
        and Fall of German Science*.

        Cited herein as item 77.

817.   Merleau-Ponty, Jacques, "L'*Essai sur la philosophie des
        sciences* d'Ampère." *Rev. Hist. Sci.* 30 (1977): 113-
        118.

        An account of Ampère's last work in which he presented
        a classification of the different sciences, demonstrating
        both his "spiritualism" and his intense scientific
        realism.

818.   Merz, John Theodore, *A History of European Thought in
        the Nineteenth Century*. 4 vols. Edinburgh, 1896-
        1914/London, 1904-1912. Reprinted, New York: Dover,
        1965.

        A monumental classic covering both scientific and
        philosophical thought in the 19th century. Although
        flawed in some details, still gives the best broad
        treatment in English of science in Germany and good
        treatments of science in the other parts of Europe.

819.   Moyer, Donald F., "The Use of Dynamics as the Basis of
        Physical Theory by British Theoretical Physicists in
        the Latter Half of the Nineteenth Century." Ph.D.
        dissertation, University of Wisconsin, 1973.

        Examines the methodology of the physicists, especially
        in relation to the energy concept, arguing that the

foundation of their work was the identification of
empirical laws with the formal relations of dynamics,
and the use of complex mechanical theories to resolve
"inherent antinomies."

820. Nedelkovitch, Duchan, "Les principaux savants yougoslaves
au XIXe siècle." *Rev. Hist. Sci.* 13 (1960): 317-323.

Lists scholars and their principal works; in the case
of the physicists, these are generally textbooks.

821. O'Hara, J.G., "George Johnstone Stoney, F.R.S., and
the Concept of the Electron." *Notes Rec. Roy. Soc.
Lond.* 29 (1974-75): 265-276.

A brief overview of the work of Stoney--best known
for his invention of the word "electron"--in molecular
physics. Topics mentioned include his early advancing
of the kinetic theory of gases; his espousal of various
"fundamental" systems of units including, in 1874, the
determination of a "definite quantity" of electricity;
and theories that spectra are generated by undulations
in the aether caused by rotating molecules (1868), or
(1891) by the orbiting of "electrons."

822. Olesko, Kathryn Mary, "The Emergence of Theoretical
Physics in Germany: Franz Neumann and the Königsberg
School of Physics, 1830-1890." Ph.D. thesis, Cornell
University, 1980. 545 pp.

An outstanding thesis based on extensive research on
previously untapped archival material. Includes some
discussion of Neumann's contribution to electromagnetic
theory, but the emphasis throughout is on wider questions
concerning the conceptualization of physics as a mathematical science and the institutionalization of the
new approach.

823. Olson, Richard, *Scottish Philosophy and British Physics
1750-1880: A Study in the Foundations of the Victorian
Scientific Style.* Princeton, N.J.: Princeton University Press, 1975. viii + 349 pp.

Argues that the methodological outlook of many leading
19th-century British physicists, especially Scottish
ones up to and including Maxwell, was formed chiefly
under the influence of the Scottish Common Sense philosophy "as taught by Thomas Reid and Dugald Stewart and
modified by Thomas Brown and Sir William Hamilton."
Reviewed by John R.R. Christie, *Ann. Sci.* 33 (1976):
311-318, and P.M. Heimann, *Isis* 67 (1976): 626-628.

824. Panter, John R., "Sir John F.W. Herschel and Scientific
      Thought in Early Nineteenth-Century Britain." Ph.D.
      dissertation, University of New South Wales (Aus-
      tralia), 1976.

      Examines Herschel's influential *Preliminary Discourse
      on the Study of Natural Philosophy* (1830), one of the
      first attempts to grapple with philosophy of science by
      a practising scientist, with an emphasis on the method
      of hypothesis and verification.

825. Paul, Harry W., "La science française de la second partie
      du XIXe siècle vue par les auteurs anglais et améri-
      cains." *Rev. Hist. Sci.* 27 (1974): 147-163.

      Questions the validity of the theories current in
      contemporary British and American historical research
      that there was an actual decline in French science
      during the period stated.

826. Pihl, Mogens, *Der Physiker L.V. Lorenz.* Copenhagen:
      Einar Munksgaard, 1939. 128 pp.

      A University of Copenhagen physics doctoral thesis.
      Includes a brief (3 pp.) biographical outline of Lorenz
      and a bibliography of his writings on physical subjects,
      together with critical analyses of his work on optics,
      the conductivity of metals, and kinetic and elasticity
      theory. Seeks throughout to clarify Lorenz' ideas by
      translating them into less complicated and more satisfac-
      tory modern mathematical form.

827. Pihl, Mogens, "The Scientific Achievements of L.V.
      Lorenz." *Centaurus* 17 (1972): 83-94.

      A brief account, in English, of Lorenz' principal
      contributions to mathematical physics, especially in
      optics, the electromagnetic theory of light, conductivity,
      and the theory of telephone cables.

828. Pihl, Mogens, "Two Contributions of L.V. Lorenz to
      Mathematical Physics." *Centaurus* 24 (1980): 361-368.

      Gives a brief account of Lorenz' solutions, as set
      out and credited to him by his friend C. Christiansen
      in his *Indledning til den matematiske Fysik* (1887/9),
      to two minor problems in mathematical physics, viz.
      (a) a stationary ellipsoid in a moving incompressible
      fluid, and (b) the rate of formation of ice.

829. Pyenson, Lewis, "Mathematics, Education and the Göttingen Approach to Physical Reality, 1890-1914." *Europa* 2 (1979): 91-127.

While chiefly concerned with post-1900 developments, the author also describes the process by which pure mathematics came to dominate German education at the expense of natural sciences in the later decades of the 19th century, contributing to a mathematical, instrumentalist view of physics.

829a. Pyenson, Lewis, "La place des sciences exactes en Allemagne à l'époque de Guillaume II." *Europa* 4 (1981): 187-217.

Surveys the social circumstances of physicists in Germany and its cultural dependencies in the period 1890-1918.

830. Quetelet, L.A.J., *Sciences mathématiques et physiques chez les Belges au commencement du XIX siècle.* Bruxelles: H.T. van Buggenhoudt, 1886. 754 pp.

A more detailed study than the author's previous work (item 441), this concentrates on the period from the late 18th century to the 1850s and links the work of Belgian scientists with that of the wider scientific community. It is divided into four sections: the first looks at the general history of science in the period; the second gives career biographies of Belgian *savants*; the third a similar treatment of Belgian writers and artists; and the last deals with foreign *savants* in their relations with Belgian science.

831. Rezneck, Samuel, "The Education of an American Scientist: Henry A. Rowland, 1848-1901." *Amer. J. Phys.* 28 (1960): 155-162.

Drawing on unpublished material, describes Rowland's education and early professional life, showing that while he took advantage of all available scientific facilities, he also engaged in extensive extra-curricular scientific work, sacrificing his income in order to buy instruments.

832. Rezneck, Samuel, "An American Physicist's Year in Europe: Henry A. Rowland, 1875-1876." *Amer. J. Phys.* 30 (1962): 877-886.

From unpublished material, briefly describes Rowland's travels, his comments on the equipment he found at the

various scientific institutions he visited, his meetings
with noted scientists such as Thomson, Maxwell and
Helmholtz, and his period of research in Helmholtz'
laboratory.

833. Schaffner, Kenneth F., ed., *Nineteenth-Century Aether
      Theories*. With a commentary by K.F. Schaffner.
      Oxford/New York: Pergamon Press, 1972. ix + 278 pp.

      A volume in the Commonwealth and International Library
      series, "Selected Readings in Physics." Comprises an
      excellent 121-page introduction by Schaffner, together
      with extracts from the writings of Fresnel, Stokes,
      Michelson and Morley, Green, MacCullagh, W. Thomson,
      FitzGerald, Heaviside, Larmor and Lorentz.

834. Sharlin, Harold I., *The Convergent Century: The Unifica-
      tion of Science in the Nineteenth Century*. London/
      New York/Toronto: Abelard-Schuman, 1966. 229 pp.

      Briefly discusses the leading themes of nineteenth-
      century physical science--electricity, heat, electro-
      magnetism, the atomic-molecular theory, wave optics,
      kinetic theory, chemical thermodynamics--focusing in
      each case on the principal publications of a few leading
      contributors to the field. The doctrine of conservation
      of energy is seen as a major unifying theme. Reviewed
      by Thomas P. Hughes, *Amer. Hist. Rev.* 72 (1967): 1346.

835. Sharlin, Harold Isadore, "William Thomson's Dynamical
      Theory: An Insight into a Scientist's Thinking."
      Ann. Sci. 32 (1975): 133-147.

      Describes Thomson's development of a unified, dynamical
      view of nature, replacing the disjoint imponderables
      of the statical view. Describes his method of contruct-
      ing first physical and then mathematical models by
      analogy, drawing upon the work of Faraday and Joule.

836. Silliman, Robert H., "William Thomson: Smoke Rings and
      Nineteenth-Century Atomism." *Isis* 54 (1963): 461-
      474.

      Traces Thomson's "smoke-ring" theory to his desire
      for mechanical models, his awareness of Helmholtz'
      hydrodynamical work on vortices, and Tait's striking
      demonstrations with smoke rings. Briefly describes
      the model and notes the widespread approval of it, but
      argues that there was never wholehearted commitment to
      it on account of its abstract mathematical nature.

837. Smith, Crosbie W., "'Mechanical Philosophy' and the Emergence of Physics in Britain, 1800-1850." *Ann. Sci.* 33 (1976): 3-29.

   Argues that a unified mathematical physics based on the energy principle was a consequence of the fusion of the Scottish "common sense" philosophy, with its idea of a mechanical philosophy in which natural phenomena were linked conceptually to dynamical laws, with the Cambridge approach with its mathematically orientated analysis of phenomena.

838. Smith, Crosbie, "Engineering the Universe: William Thomson and Fleeming Jenkin on the Nature of Matter." *Ann. Sci.* 37 (1980): 387-412.

   Examines the intellectual background to, and content of, the unpublished correspondence in 1867 between Thomson and Jenkin on the nature of matter, using this to point out, *contra* Doran (item 1132), the limitations inherent in inventing intellectual traditions for ideas without fully accounting for their immediate historical background.

839. Snelders, H.A.M., "A.M. Mayer's Experiments with Floating Magnets and Their Use in the Atomic Theories of Matter." *Ann. Sci.* 33 (1976): 67-80.

   Describes Alfred Marshall Mayer's experiments with floating magnets in 1878 and 1879, and discusses how his results were used by Kelvin and J.J. Thomson as models for their respective theories of atomic structure.

840. Sutton, M.A., "J.F. Daniell and the Boscovichean Atom." *Studs. Hist. Phil. Sci.* 1 (1971): 277-292.

   From an examination of the works of Faraday's friend Daniell, Sutton argues that there was no overt hostility in the 1830s to the Boscovichean theory, but also concludes that neither Daniell nor Faraday was particularly convinced by the theory.

841. Thiele, Joachim, "Aus der Korrespondenz Ernst Machs: Briefe deutscher und englischer Naturwissenschaftler." *NTM* 7(1) (1970): 66-75.

   Publishes nine letters, including one of Mach's to Kirchhoff and eight to him from Kirchhoff, Helmholtz, Ludwig, Tyndall, Kelvin and others.

842. Thiele, Joachim, "Zur Wirkungsgeschichte des Doppler-
     prinzips im Neunzehnten Jahrhundert." *Ann. Sci.* 27
     (1971): 393-407.

     An account of the formulation of the concept of the
     Doppler Shift and its application in physics, especially
     in the fields of acoustics and optics. Discusses the
     work of Buys-Ballot, Mach, Fizeau, Huggins and Vogel.

843. Topper, D.R., "'To Reason by Means of Images': J.J.
     Thomson and the Mechanical Picture of Nature." *Ann.
     Sci.* 37 (1980): 31-57.

     A discussion of the consistently mechanical physical
     theories, based on the Lagrangian formulation of dy-
     namics, used by J.J. Thomson throughout his professional
     life.

844. Williams, L. Pearce, "The Physical Sciences in the
     First Half of the Nineteenth Century: Problems and
     Sources." *Hist. Sci.* 1 (1962): 1-15.

     Calls for further research into three fundamental
     areas of matter theory, 1800-1850, namely: the nature
     and action of imponderable fluids; the interaction of
     these fluids with ponderable matter; and the ultimate
     constitution of ordinary matter. Also stresses the
     need to examine manuscript material to discover the
     true genesis of ideas, given the artificially "inductive"
     presentation of formal scientific publications.

845. Wilson, D.B., "The Reception of the Wave Theory of Light
     by Cambridge Physicists (1820-50): A Case Study in the
     Nineteenth-Century Mechanical Philosophy." Ph.D.
     dissertation, Johns Hopkins University, 1968.

     Using as an example the introduction and development
     of the wave theory of light by Herschel, Whewell, Airy,
     Challis and Stokes, demonstrates the domination of the
     mechanical world view in Cambridge during the period in
     question, as shown in the status accorded to mechanical
     laws, in the insistence on the necessity of an ether to
     provide a mechanism for light, and in the assumption
     that the microscopic and macroscopic worlds followed
     identical mechanical processes.

846. Wilson, David B., "Herschel and Whewell's Version of
     Newtonianism." *J. Hist. Ideas* 35 (1974): 79-97.

     Suggests that while Herschel and Whewell disagreed
     on the status of Newton's laws--Herschel being a thorough-

going empiricist and Whewell appealing to the necessary truth of the laws as "fundamental" ideas--they both proposed a well-defined Newtonian methodology whereby the use of reason, analogy and experiment could uncover true causes.

847. Wilson, D.B., "Kelvin's Scientific Realism: The Theological Context." *Phil. J.* 11 (1974): 41-60.

Argues that Kelvin adopted a view of mind and knowledge that was widely held in Glasgow and Cambridge during his student days in the 1830s and '40s, in which the theological argument from design played a central role. Suggests that this underlay his life-long conviction that man's knowledge of unobservable entities was genuine.

848. Wilson, David B., ed., *Catalogue of the Manuscript Collections of Sir George Gabriel Stokes and Sir William Thomson, Baron Kelvin of Largs, in the Cambridge University Library.* Cambridge: Cambridge University Library, 1976. Separately paginated: 589 pp. + 363 pp.

An invaluable guide to two large and important collections of manuscripts.

848a. Wilson, David B., "Concepts of Physical Nature: John Herschel to Karl Pearson." *Nature and the Victorian Imagination*, ed. V.C. Knoepflmacher and G.B. Tennyson. Berkeley: University of California Press, 1977. pp. 201-215.

Surveys the changes that took place during the Victorian era in the attitudes of British natural philosophers towards physical nature, our knowledge of it, and God's place in the scheme of things.

848b. Wise, M. Norton, "The Maxwell Literature and British Dynamical Theory." *Hist. Studs. Phys. Sci.* 13 (1982): 175-205.

An essay review of the large body of recent writings on Maxwell. Identifies as a potentially unifying thread the question of the status of geometry in 19th-century British formulations of mathematical physics.

849. Wynne, Brian, "Physics and Psychics: Science, Symbolic Action and Social Control in Late Victorian England." *Natural Order: Historical Studies of Scientific Culture*, ed. Barry Barnes and Steven Shapin. Beverly Hills/London: Sage, 1979, pp. 167-186.

Argues that the "Cambridge school" of physicists was attracted to anti-positivist metaphysical theories such

as the etherial basis of matter, and even spiritualism,
as a reaction against the "new" utilitarian, anti-
metaphysical and materialist view of science increasingly
prevalent at the time.

(b) Institutional Histories

850.   Berman, Morris, *Social Change and Scientific Organiza-*
       tion: The Royal Institution, 1799-1844.   Ithaca, N.Y.:
       Cornell University Press, 1978. xxviii + 224 pp.

Sees the Royal Institution as a focus for changes by
which science became the "ideology" for industrialized
society. Argues that the work of Davy, Faraday and
others for the Institution exemplified the increasing
role of science in technology, the use of science as
the model for a utilitarian society, and the change from
amateur to professional scientist, all of which served
to promote the ideal of an industrial society.

851.   Brock, W.H., "The Spectrum of Scientific Patronage."
       The Patronage of Science in the Nineteenth Century,
       ed. G.L'E. Turner. Leiden: Noordhoff International,
       1976, pp. 173-206.

Describes the many different private, institutional
and governmental positions in 19th-century Britain that,
though often non-research and largely administrative in
character, required technical training of their incum-
bents and thus provided scientists with a means of
earning a livelihood.

852.   Buchheim, Gisela, "Initiativen zur Gründung der Physi-
       kalisch-Technische Reichsanstalt (1887)." *NTM* 11(2)
       (1974): 33-43.

Describes the events leading up to the establishment
of the Physikalisch-Technische Reichsanstalt, emphasizing
the contribution of Werner von Siemens and the industrial
role envisaged for the institute.

853.   Buchheim, Gisela, "Reichstagsdebatten über die Gründung
       der Physikalisch-Technische Reichsanstalt." *NTM* 12(2)
       (1975): 1-13.

Describes the political debates over the founding of
the Physikalisch-Technische Reichsanstalt, arguing that
it was intended to be primarily a standards laboratory
to meet the needs of German industry.

854. Buchheim, Gisela, "Die Entwicklung des elektrischen
     Messwesens und die Gründung der Physikalisch-Technische
     Reichsanstalt." *NTM* 14(1) (1977): 16–32.

     Discusses the role envisaged for the Physikalisch-
     Technische Reichsanstalt in the field of electrical
     measurement standards.

855. Buck, Barbara Reeves, "Italian Physicists and Their
     Institutions, 1861–1911." Ph.D. thesis, Harvard Uni-
     versity. 1980. 758 pp.

     A very thorough study that concentrates on the organi-
     zational framework within which physics was cultivated
     in Italy in the half-century following the political
     unification of the country.

856. Cahan, David, "Werner Siemens and the Origin of the
     Physikalisch-Technische Reichsanstalt, 1872–1887."
     Hist. Studs. Phys. Sci. 12 (1982): 253–283.

     Argues with extensive archival documentation that
     the major drive for the establishment of the PTR came
     from "Siemens' personal desires for the advancement of
     pure science, the securing of scientific foundations
     for technology, and his patriotism."

857. Cardwell, Donald S.L., *The Organization of Science in
     England*. Revised ed. London: Heinemann, 1972.
     xii + 268 pp.

     A broad treatment of the scientific institutions of
     19th-century England. Discusses the development of a
     system of scientific and technical education, including
     the controversies leading to the Great Exhibition of
     1851, the "stunted" growth of the mechanics' institutes,
     the reform of Oxford and Cambridge, and the rise of the
     new universities. Reviewed by A.D. Orange, *Brit. J.
     Hist. Sci.* 7 (1974): 298–299.

858. [Cavendish Laboratory], *A History of the Cavendish
     Laboratory, 1871–1910*. London: Longmans, Green & Co.,
     1910. xii + 342 pp.

     A publication marking the 25th anniversary of J.J.
     Thomson's taking up the Cavendish Professorship of
     Experimental Physics at Cambridge. Comprises nine
     chapters by different authors surveying the research
     and teaching activities of the Laboratory, together
     with a list of research publications and another of
     "those who have worked in the Laboratory" to that date.

859. Crosland, Maurice, "The French Academy of Sciences in
     the Nineteenth Century." *Minerva* 16 (1978): 73–102.

Presents valuable information on, *inter alia*, the
structure of the Academy and the differentiation of
scientific disciplines within it, the procedures for
electing new members, the intellectual and political
divisions among the members, their sources of income,
and the role of the Academy in 19th-century French
scientific life.

860. Crowther, J.G., *The Cavendish Laboratory, 1874-1974*.
       London: Macmillan, 1974. xvi + 464 pp.

       A narrative history marking the centenary of the
       Laboratory. Includes information on the background
       to the establishment of the Laboratory and several
       chapters on its development under its 19th-century
       directors, Maxwell, Rayleigh and J.J. Thomson.

861. Foote, George A., "The Place of Science in the British
       Reform Movement 1830-1850." *Isis* 42 (1951): 192-208.

       Describes the criticisms of British science, and of
       the lack of government support, by authors such as John
       Herschel, Babbage and Bulwer-Lytton, and also some re-
       sponses to these criticisms.

862. Forman, Paul, John L. Heilbron and Spencer Weart, "Physics
       circa 1900: Personnel, Funding and Productivity of the
       Academic Establishments." *Hist. Studs. Phys. Sci.* 5
       (1975): 1-185.

       Drawing from a wide range of material (listed in an
       exhaustive bibliography), presents a geographical survey
       of institutional academic physics *c.* 1900.

863. Fox, Robert, "Scientific Enterprise and the Patronage
       of Research in France, 1800-1870." *Minerva* 11 (1973):
       442-473. Reprinted, *The Patronage of Science in the
       Nineteenth Century*, ed. G.L'E. Turner. Leiden:
       Noordhoff International, 1976, pp. 9-51.

       Argues that government patronage of science in France
       did not significantly decline after the Bourbon Restora-
       tion. The decline of research achievement in France is
       to be attributed, rather, to the lack of initiative on
       the part of individual scientists and to the changing
       style of Restoration intellectual life which tended to
       favor a generalized approach to science with an em-
       phasis on lucrative public lecturing skills, rather
       than specialized research.

864. Fox, Robert, and George Weisz, eds., *The Organization of Science and Technology in France, 1808-1914.* Cambridge: Cambridge University Press/Paris: Editions de la Maison des Sciences de l'Homme, 1980. x + 355 pp.

   A valuable collection of essays which treat the institutions for both education and research in science and technology in 19th-century France. Several contributors question the traditional view that French science declined during the period. Reviewed by Mary Jo Nye, *Isis* 73 (1982): 144-145.

865. Hays, J.N., "Science and Brougham's Society." *Ann. Sci.* 20 (1964): 227-241.

   Examines the publications on natural philosophy of the Society for the Diffusion of Useful Knowledge (1826-1846) aimed at educating the broader British public. Concludes that they were characterized by the espousal of the themes of "utility, basic empiricism, belief in a Providential Cosmos and a faith in the possibility and desirability of broad scientific education."

866. Jungnickel, Christa, "Teaching and Research in the Physical Sciences and Mathematics in Saxony, 1820-1850." *Hist. Studs. Phys. Sci.* 10 (1979): 3-48.

   Traces the rise of the university (especially Leipzig University) as a place for research and training in research work. Discusses, with examples, the evolution of a consensus concerning a research methodology combining rigorous collaborative experimentation and theorizing. Finally, discusses the foundation of the Saxon Society of Sciences, which is seen as embodying the Saxons' awareness of themselves as a research-orientated scientific community.

867. MacLeod, Roy M., "Resources of Science in Victorian England: the Endowment of Science Movement, 1868-1900." *Science and Society, 1600-1900*, ed. Peter Mathias. Cambridge: Cambridge University Press, 1972, pp. 111-166.

   A study of changing social attitudes towards scientific research in Victorian England.

868. MacLeod, Roy M., "Science and the Treasury: Principles, Personalities and Policies, 1870-1885." *The Patronage*

of Science in the Nineteenth Century, ed. G. L'E.
Turner. Leiden: Noordhoff International, 1976, pp.
115-172.

Provides an account of the extent of British government funding of science and some of the reasons for
its apparent restrictiveness, including the doubtful
financial rewards of such support and the bureaucratic
nature of the Treasury.

869. MacLeod, Roy, and Peter Collins, *The Parliament of
Science: The British Association for the Advancement
of Science, 1831-1981.* Northwood, Middlesex: Science
Reviews Ltd., 1981. vi + 308 pp.

A collection of essays marking the 150th anniversary
of the Association. The following discuss 19th-century
developments: "Retrospect: The British Association and
Its Historians" and "Introduction: On the Advancement
of Science" by Roy MacLeod; "The Beginnings of the
British Association, 1831-1851" by A.D. Orange; "Scien-
tific Method and the Image of Science, 1831-1890" by
Richard Yeo; "Advancing Science: The British Association
and the Professional Practice of Science" by W.H. Brock;
"The British Association and the Provincial Public" by
Philip Lowe; "Scientific Internationalism and the
British Association" by Giuliano Pancaldi; "The British
Association and Empire: Science and Social Imperialism,
1880-1940" by Michael Worboys; and "The Schooling of
Science in England, 1854-1939" by David Layton.

870. MacLeod, Roy, and Russell Moseley, "The 'Naturals' and
Victorian Cambridge: Reflections on the Anatomy of
an Elite, 1851-1914." *Oxford Rev. Educ.* 6 (1980):
177-195.

A survey of the impact of the new Natural Sciences
Tripos on Cambridge educational patterns during the
second half of the 19th century.

871. Morrell, J.B., "Individualism and the Structure of
British Science in 1830." *Hist. Studs. Phys. Sci.* 3
(1971): 183-204.

Argues that, despite widespread local disillusionment
with British science and approval of State-organized
French science, the dominant individualistic tradition of
British science was maintained through the Age of Reform,
even in the British Association for the Advancement of

Science which "quietly and quickly" dropped the aim of national encouragement of science.

872. Morrell, J.B., "The Patronage of Mid-Victorian Science in the University of Edinburgh." *Science Studies* 3 (1973): 353-388. Reprinted, *The Patronage of Science in the Nineteenth Century*, ed. G.L'E. Turner. Leiden: Noordhoff International, 1976, pp. 53-93.

Using Edinburgh as an example, demonstrates the almost total lack of government funding of university-based research in mid-Victorian Britain. In general any such work had to be sponsored by the university, by patrons, or, as in the case of P.G. Tait, out of the researcher's own pocket.

873. Morrell, Jack, and Arnold Thackray, *Gentlemen of Science: Early Years of the British Association for the Advancement of Science*. Oxford: Clarendon Press, 1981. xxiv + 592 pp.

A detailed study, based on a wide range of manuscript and printed sources, of the scientific milieu in Britain in the 1830s and '40s.

874. Moseley, Russell, "Tadpoles and Frogs: Some Aspects of the Professionalization of British Physics, 1870-1939." *Soc. Studs. Sci.* 7 (1977): 423-446.

Focuses on the "rank and file physicist," working at some distance from the frontiers of research, as an important aspect of the professionalization of physics in Britain. Much of the discussion concerns 20th-century developments, but some account is also given of the establishment of the Physical Society of London in 1874 and its subsequent activities.

875. Pfetsch, Frank, "Scientific Organization and Science Policy in Imperial Germany, 1871-1914: The Foundation of the Imperial Institute of Physics and Technology." *Minerva* 8 (1970): 557-580.

Attempts a description of "the fundamental features of science policy in imperial Germany," using the foundation of the Physikalisch-Technische Reichsanstalt as a case study. Is eventually driven to the conclusion, however, that the idea of the Institute arose entirely independently of "such cultural or science policy-making institutions as then existed."

876.   Phillips, Melba, "Early History of Physics Laboratories
       for Students at the College Level." *Amer. J. Phys.*
       49 (1981): 522-527.

       A brief review of the introduction of student labora-
       tory classes in physics teaching *c.* 1870, especially by
       E.C. Pickering at the Massachusetts Institute of Tech-
       nology.

877.   Pyenson, Lewis, and Douglas Skopp, "Educating Physicists
       in Germany *circa* 1900." *Soc. Studs. Sci.* 7 (1977):
       329-366.

       A demographic survey of Ph.D. recipients in German
       universities in the years around 1900, based on a
       sample of 281 doctoral dissertations.

878.   Shinn, Terry, "The French Science Faculty, 1808-1914:
       Institutional Change and Research Potential in Mathe-
       matics and the Physical Sciences." *Hist. Studs. Phys.*
       Sci. 10 (1979): 271-332.

       By demographic analysis shows that neither centraliza-
       tion nor regionalization of authority in 19th-century
       French science faculties determined research productivity.
       Both development and stagnation occurred in different
       times and places, but this was due to other factors,
       such as the greater or lesser emphasis on teaching and
       industrial work decreed by the government, funding,
       individual flair, level of commitment to research, and
       the higher or lower research standards required to
       obtain qualifications.

879.   Sviedrys, Romualdas, "The Rise of Physical Science at
       Victorian Cambridge." With commentary by Arnold
       Thackray and reply by Romualdas Sviedrys. *Hist.*
       Studs. Phys. Sci. 2 (1970): 127-151.

       Argues that there was a twenty- to thirty-year
       "generation gap" for reforms at 19th-century Cambridge,
       including those of science, and that the Cavendish
       Laboratory was an exception mainly due to its external
       funding. Thackray suggests that this argument should
       be extended to all the "old" institutions in Britain.

880.   Sviedrys, Romualdas, "The Rise of Physics Laboratories
       in Britain." *Hist. Studs. Phys. Sci.* 7 (1976): 405-436.

       Shows that initially during the 19th century labora-
       tories were small and private, and that as they became

officially recognized or established, they also became increasingly tied to resolving industrial problems and to the elementary training of engineers, civil servants and teachers. By the 1880s, there was an increasing separation and professionalization of engineering and academic physics. This culminated in the work of the Cavendish Laboratory, unique as a well-funded graduate research laboratory, whose influence came to dominate British physics.

881. Thompson, D., "John Tyndall and the Royal Institution." *Ann. Sci.* 13 (1957): 9-22.

Describes Tyndall's successful style as a lecturer and shows how he contributed to the popularization of science. Includes Tyndall's notes on a lecture on electricity.

882. Turner, R. Steven, "The Growth of Professorial Research in Prussia, 1818 to 1894--Causes and Context." *Hist. Studs. Phys. Sci.* 3 (1971): 137-182.

Argues that the unique domination of research rather than teaching in the Prussian universities was due not only to the glorification of *Wissenschaft* and the stimulus of competition between universities, but also to the government's active adoption of these values, encouraging competition and promoting individuals primarily on the basis of success in research.

883. Wilson, David B., "Experimentalists among the Mathematicians: Physics in the Cambridge Natural Sciences Tripos, 1851-1900." *Hist. Studs. Phys. Sci.* 12 (1982): 325-371.

Includes much detail on the structure and content of both the Mathematics Tripos and the Natural Sciences Tripos at 19th-century Cambridge, and on the career patterns of the most successful graduates from the two programs. Charts the changing relationship between the programs, especially the increasing prestige of the Natural Sciences Tripos toward the end of the period.

(c) Instrumentation

883a. d'Agostino, S., and M.G. Ianniello, "Elettrometri e galvanometri dell' Ottocento: evoluzione strumentale e contesti teorici." *Annali dell'Istituto e Museo di Storia della Scienza di Firenze* 5(2) (1980): 69-82.

Describes and places in historical context some of the 19th-century electrical instruments in the museum of the Institute of Physics, University of Rome.

*   Auerbach, F., *Ernst Abbe: Sein Leben, sein Wirken, seine Personlichkeit.*

Cited herein as item 31.

884.   Brock, W.H., "William Prout and Barometry." *Notes Rec. Roy. Soc. Lond.* 24 (1969): 281-294.

Describes Prout's "expensive and accurate" barometer and also a similar one, the construction of which he supervised for the Royal Society. Also mentions some of the results achieved with this instrument, including an accurate density determination for air and the discovery that the mercury and air thermometers were not fully complementary.

885.   Brown, Sanborn C., "Discovery of the Differential Thermometer." *Amer. J. Phys.* 21 (1953): 13-17.

Describes the acerbic priority dispute between Rumford and Sir John Leslie over the invention of the differential thermometer in 1803 and another equally violent dispute between Davy (on behalf of van Helmont) and Leslie in 1812-13.

885a. Frison, E., *L'évolution de la partie optique du microscope au cours du dix-neuvième siècle: les test objets, les test-, probe- et typen platten.* Leiden: Rijksmuseum voor de Geschiedenis der Natuurwetenschappen, 1954. 168 pp.

Not seen.

886.   Green, George, and J.T. Lloyd, *Kelvin's Instruments and the Kelvin Museum.* Glasgow: University of Glasgow Press, 1970. 68 pp.

A description and discussion of the historical significance of apparatus in the museum including large numbers of instruments upon which Kelvin performed his electrical measurements as well as those used by Joule, Stirling, and Kerr. Also describes Kelvin's courses and teaching methods. Illustrated by 20 plates.

886a. Körber, Hans-Günther, "Aus der Entwicklung der Bimetall-
       thermometer zu Anfang des 19. Jahrhunderts." *NTM*
       (1964): Beiheft, 102-107.

       A brief account of early work on bimetallic thermome-
       ters, including a description of an exemplar in the
       Hellmann Collection of the Geomagnetic Institute, Potsdam.

887.   Nuttall, R.H., and A. Frank, "Makers of Jewel Lenses in
       Scotland in the Early Nineteenth Century." *Ann. Sci.*
       30 (1973): 407-416.

       Shows that there was a group of makers of jewel micro-
       scopes in Scotland in the 1840s, and describes from the
       limited information available the work and materials of
       four of them--Adie, Blackie, Veitch and Hill--whose
       products were used by Brewster.

888.   Ronchi, Vasco, "Giovan Battista Amici's Contribution to
       the Advances of Optical Microscopy." *Physis* 11 (1969):
       520-533.

       Describes Amici's development towards the middle of
       the 19th century of improved objectives which led to a
       dramatic increase in the resolving power of achromatic
       microscopes. Also includes modern analyses, due to
       P.H. Van Cittert and J.G. Van Cittert-Eymers and pre-
       viously published by them, of the resolving power of
       various 18th and 19th-century microscopes.

889.   Tsuneishi, Kei-ichi, "On the Abbe Theory (1873)."
       Japanese Studs. Hist. Sci. 12 (1973): 79-92.

       Compares Abbe's theory of the optics of the microscope
       with that of Helmholtz (1874) showing that it was more
       comprehensive, even though the latter was initially the
       more popular.

890.   Tsuneishi, Kei-ichi, "On Stoney's Concept of Image,
       with Reference to the Formalization of the Abbe
       Theory." *Japanese Studs. Hist. Sci.* 14 (1975): 95-
       102.

       Briefly discusses Stoney's formalization (1895), based
       on the electromagnetic theory of light, of Abbe's dif-
       fraction method for analyzing optical instruments.

(d) Foundational Issues,
Properties of Matter, Sound

891. Bikerman, J.J., "Theories of Capillary Attraction."
      Centaurus 19 (1975): 182-206.

      Summarizes, compares and assesses the work of Laplace
      (1805), Young (1805), Poisson (1831), Neumann (1864),
      and Dupré (1866).

892. Bork, Alfred M., "The 'FitzGerald' Contraction." *Isis*
      57 (1966): 199-207.

      Argues that while FitzGerald seems to have had the
      general idea that the null result of the Michelson-
      Morley experiment could be explained by a contraction
      in length in the direction of motion, and seems to
      have mentioned it to Lodge and Lorentz, he published
      nothing. The attribution of the idea to him seems to
      have resulted from reports made by friends after his
      death. (For new evidence contradicting this argument,
      see the article by Brush, item 893.)

*   Brouzeng, Paul, "Poisson et la capillarité selon Duhem
      d'après un manuscrit inédit: les leçons sur les
      théories de la capillarité."

      Cited herein in item 742.

893. Brush, S.G., "Note on the History of the FitzGerald-
      Lorentz Contraction." *Isis* 58 (1967): 230-232.

      Shows that FitzGerald in fact did have a brief letter
      published in 1889, though without his knowledge, in
      which was set out his idea of a contraction in length
      that might explain the null result of the Michelson-
      Morley experiment. (Cf. the article by Bork, item 892.)

894. De Kosky, Robert Keith, "The Scientific Work of Sir
      William Crookes." Ph.D. dissertation, University of
      Wisconsin, 1972.

      Discusses both Crookes' chemical and his physical
      work, especially the radiometer phenomena and his
      "fourth state of matter" theory.

895. De Kosky, Robert K., "William Crookes and the Fourth
      State of Matter." *Isis* 67 (1976): 36-60.

      Shows how Crookes' brilliant experiments with radio-
      meters and cathode ray tubes were, from 1876 onwards,

dominated and determined by his conception of a fourth
state of matter beyond the gaseous state, occurring
in near-vacuum conditions where random molecular movement was transformable into directed streaming normal
to a heated or charged surface. These ideas were received with almost universal skepticism since alternative
explanations, more compatible with kinetic theory, were
available.

* Finn, Bernard S., "Laplace and the Speed of Sound."

Cited herein as item 723.

896. French, A.P., "Earliest Estimates of Molecular Size."
     Amer. J. Phys. 35 (1967): 162-163.

Briefly describes Young's (1816) and Waterston's
(1858) methods of attaining their estimates.

897. Holton, Gerald, "Einstein, Michelson and the 'Crucial'
     Experiment." *Isis* 60 (1969): 133-197.

Demonstrates convincingly that Einstein's theory and
Michelson's experiment were connected only retrospec-
tively. Also, however, shows the hold which the concept of the aether had on the scientific world at the
time, and brings out well the puzzlement aroused by
the null result obtained in the experiment.

898. Hopley, I.B., "Clerk Maxwell's Apparatus for the Measure-
     ment of Surface Tension." *Ann. Sci.* 13 (1957): 180-
     187.

Discusses an unpublished manuscript of Maxwell's and
the apparatus described in it, some of which still sur-
vives.

899. Kittel, C., "Larmor and the Pre-History of the Lorentz
     Transformations." *Amer. J. Phys.* 42 (1974): 726-729.

Suggests that Larmor's anticipation of the Lorentz
transformations, published in 1900, was neglected due
to his failure to express his idea clearly and to his
own lack of awareness of the significance of his work.

* Kuhn, Thomas S., "The Caloric Theory of Adiabatic Com-
  pression."

Cited herein as item 1019.

900. Newburgh, Ronald, "Fresnel Drag and the Principle of
      Relativity." *Isis* 65 (1974): 379-386.

      Describes the experiments of Mascart and the theoretical
      work of Potier and Veltmann (1870-74) explaining, in
      terms of another theory, why first-order (v/c) effects
      of an aether wind entrained by the earth were not ob-
      servable, and leading to the search for second-order
      (v^2/c^2) effects by Michelson and others.

901. Pas, Peter W. van der, "The Early History of Brownian
      Motion." *Proceedings, 12th International Congress
      of the History of Science, Paris, 1968* (Paris, 1971),
      vol. 8, pp. 143-158.

      Details Brown's observations and his conclusion that
      the motion was not organic in nature (1828). Also
      lists precursors including Desaguliers, Boerhaave,
      Stephen Gray and Ingenhousz.

902. Semmel, Bernard, "Parliament and the Metric System."
      Isis 54 (1963): 125-133.

      Describes the attempts to introduce the metric system
      into England in the early 19th century and in the 1860s,
      and suggests reasons for its rejection. Semmel's article
      is criticized by Joseph Mayer, "Parliament and the
      Metric System--Comments," *Isis* 57 (1966): 117-119, with
      a reply by Semmel, *Isis* 57 (1966): 119-120.

903. Swenson, Loyd S., Jr., *The Etherial Aether: A History
      of the Michelson-Morley-Miller Aether-Drift Experiments,
      1880-1930*. Austin: University of Texas Press, 1972.
      xxii + 361 pp.

      An excellent account of the aether-drift experiments
      of the American trio in their historical setting.
      Describes a large variety of aether theories, together
      with related experimental work including Hertz' discovery
      of radio waves. A good bibliography. Reviewed by Joan
      Bromberg, *Isis* 64 (1973): 431-432 and D.B. Wilson,
      Hist. Sci. 12 (1974): 220-227.

904. Swenson, Loyd S., Jr., *Genesis of Relativity: Einstein
      in Context*. New York: Burt Franklin, 1979. xvi +
      266 pp.

      Provides a summary of physics from the 1870s to the
      culmination of Einsteinian relativity. Included are
      descriptions of the relevant aspects of Maxwell's work,

the development of Maxwell's theory especially by Hertz, and the pre-Einsteinian attempts at synthetic theories, involving concepts of aether, electrons and atoms, by Hertz, Larmor, Lorentz, Poincaré and others.

905. Tabor, David, "A propos du frottement de roulement: une controverse oubliée." *Rev. Hist. Sci.* 14 (1961): 13-18.

Concerns mainly the work of Morin and Dupuit (1838-1842) though earlier and later contributions by, respectively, Coulomb and Reynolds, are also mentioned.

* Thiele, Joachim, "Zur Wirkungsgeschichte des Dopplerprinzips im Neunzehnten Jahrhundert."

Cited herein as item 842.

906. Trevena, David H., "Marcelin Berthelot's First Publication in 1850, on the Subjection of Liquids to Tension." *Ann. Sci.* 35 (1978): 45-54.

Describes Berthelot's initial paper on tension in liquids, the experiment and his conclusions; and relates this work to 20th-century work in the same field.

907. Trevena, David H., "Reynolds on the Internal Cohesion of Liquids." *Amer. J. Phys.* 47 (1979): 341-345.

Describes and analyzes Osborne Reynolds' work, *c.* 1880, and briefly compares it to the similar work of Berthelot (1850) and Donny (1846).

908. Trevena, David H., "The Pioneer Work of François Donny on the Existence of Tension in Liquids." *Ann. Sci.* 37 (1980): 378-386.

Shows that Donny's work on tension in liquids and 'cavitation' (1846) antedated Berthelot's by four years, and compares his work with that of Berthelot and Reynolds in the same field.

909. Turner, R. Steven, "The Ohm-Seebeck Dispute, Hermann von Helmholtz, and the Origins of Physiological Acoustics." *Brit. J. Hist. Sci.* 10 (1977): 1-24.

Describes Ohm's "law" of auditory perception, the controversy with Seebeck that forced him to withdraw it, and its revival and extension by Helmholtz who, by isolating those parts of the phenomena of physiological acoustics that were amenable to mechanistic principles, was able to develop the theory mathematically.

910.   Woodruff, A.E., "William Crookes and the Radiometer." *Isis* 57 (1966): 188-198.

A brief account of Crookes' discovery and development of the radiometer and his attempts to explain its action, firstly by radiation pressure, and then by a kinetic gas theory that led him to postulate an ultra-gaseous "fourth state of matter."

(e) Mechanics, Fluid Mechanics

911.   Acloque, Paul, *Oscillation et stabilité selon Foucault: critique historique et expérimentale*. Paris: C.N.R.S., 1981. xiv + 149 pp.

An account of Foucault's development of his famous pendulums and the discussions these engendered, together with Acloque's own experiments on the subject.

912.   Apmann, Robert P., "A Case History in Theory and Experiment: Fluid Flow in Bends." *Isis* 55 (1964): 427-434.

Illustrates a methodological split in 19th-century hydraulics by the reception given to two near-simultaneous solutions to the same problem. One solution, by James Thomson, was based on experiment, was not mathematically expressed, and gained wide acclaim. The other, by Boussinesq, was based on fluid theory, was highly mathematical, and remained almost totally unknown.

*   Arnold, David H., "Poisson and Mechanics."

Cited herein in item 742.

913.   Bork, Alfred M., "'Vectors Versus Quaternions': The Letters in *Nature*." *Amer. J. Phys.* 34 (1965): 202-211.

Summarizes the various arguments mounted by Gibbs, Tait, Heaviside and others between 1880 and 1900 over the relative merits of vectors and quaternions.

914.   Brillouin, Marcel, "Questions d'hydrodynamique." *Annales de la Faculté des Sciences de Toulouse pour les sciences mathématiques et les sciences physiques* 1 (1887): 1-80.

A comprehensive account of fluid flow theory in the period 1820 to 1886.

*   Bucciarelli, Louis L., "Poisson and the Mechanics of Elastic Surfaces."

Cited herein in item 742.

915.   Bucciarelli, Louis L., "Poisson, Navier and the Vibration of Elastic Surfaces: A Contrast in Style." *Proceedings, 16th International Congress of the History of Science, Bucharest, 1981* (Bucharest, 1981), vol. 1, pp. 345-350.

Suggests that a useful distinction can be made between similar work by individuals operating in different styles. In this case, argues that Poisson primarily was interested in the analytical consequences of a theoretical model while Navier emphasized the value of the method he employed and stressed the pragmatic consequences of his work.

916.   Bucciarelli, Louis L., and Nancy Dworsky, *Sophie Germain: An Essay in the Theory of Elasticity.* Dordrecht/Boston: D. Reidel, 1980. xi + 147 pp.

Focuses primarily on the developments leading to Germain's receiving the Paris Academy's *prix extraordinaire* in 1816 for her study of the vibrations of elastic surfaces. Assesses the strengths and the weaknesses of her work, giving a sensitive account of the problems she experienced, as a woman and therefore an outsider to the scientific community of her day, in pursuing research at the highest level. Reviewed by J.J. Cross, *Ann. Sci.* 39 (1982): 85-88.

917.   Cannon, Walter F., "William Whewell, F.R.S. (1794-1866). II: Contributions to Science and Learning." *Notes Rec. Roy. Soc. Lond.* 19 (1964): 176-191.

Summarizes the contents of Whewell's textbooks, noting his introduction of the techniques of the continental mathematicians into mechanics even as he continued to insist on the primacy of physical reasoning, and also his early introduction of the ideas of the French theoretical engineers, including the concept of work. Includes a brief summary of Whewell's ideas on scientific methodology.

917a. Charlton, T.M., "Maxwell, Jenkin and Cotterill and the
      Theory of Statically-Indeterminate Structures."
      Notes Rec. Roy. Soc. Lond. 26 (1971): 233-246.

      Discusses Maxwell's 1864 paper on the theory of struc-
      tures, together with subsequent refinements by Jenkin,
      who introduced the principle of virtual work into Max-
      well's method, and Cotterill, who developed the principle
      that eventually became known as the "principle of least
      work."

918. Clebsch, A., *Théorie de l'élasticité des corps solides*,
      trans. by Barré de Saint-Venant and Flamant, with
      notes by Saint-Venant. Paris, 1883. xxii + 900 pp.

      Selective historical and explanatory notes are inter-
      spersed throughout this translation of Clebsch's im-
      portant text.

919. Crowe, Michael J., *A History of Vector Analysis: The
      Evolution of the Idea of a Vectorial System.* Notre
      Dame/London: University of Notre Dame Press, 1967.
      xviii + 270 pp.

      Includes extensive discussions of the contributions
      of Hamilton, Grassmann, Tait, Maxwell, Gibbs, Heaviside
      and others. As indicated by the subtitle, the emphasis
      throughout is on the concept of a vectorial system
      rather than on the history of particular theorems in
      vectorial analysis.

920. Dobrovolsky, W., "Développement de la théorie des vec-
      teurs et des quaternions dans les travaux des mathé-
      maticiens russes du XIXe siècle." *Rev. Hist. Sci.* 21
      (1968): 345-349.

      A brief description of the contribution of major
      Russian workers to this field.

*   Gillispie, Charles C., and A.P. Yushkevich, *Lazare
      Carnot, Savant.*

      Cited herein as item 732.

921. Grigor'ian, A.T., "M.W. Ostrogradski und die Geschichte
      der theoretischen Mechanik in Russland in der ersten
      Hälfte des 19. Jahrhunderts." *Sowjetische Beiträge
      zur Geschichte der Naturwissenschaft*, ed. Gerhard
      Harig. Berlin: VEB Deutscher Verlag der Wissenschaften,
      1960, pp. 192-202.

Surveys the important contributions of Ostrogradsky to theoretical mechanics, especially in further developing Lagrange's analytical methods, and also, much more briefly, the work of the school that developed around him.

922. Grigor'ian, A.T., *Ocherki istorii mekhaniki v Rossii.* (Notes on the History of Mechanics in Russia). Moscow: Akademiya Nauk SSSR, 1961. 292 pp.

Discusses all aspects of Russian contributions to mechanics, concentrating on the 19th and 20th centuries.

923. Grigor'ian, A.T., "On the Development of Variational Principles of Mechanics." *Arch. Int. Hist. Sci.* 18 (1965): 23-35.

Describes the modifications and developments of the "principle of least action," especially in the work of Lagrange, Hamilton and Ostrogradsky.

924. Grigor'ian, A.T., "Einige Entwicklungsprobleme der technischen Mechanik in Russland in der zweiten Hälfte des 19. und Anfang des 20. Jahrhunderts." *NTM* 7(1) (1970): 23-32.

Describes some of the work of Chebyshev, Krylov and others on topics of theoretical applied mechanics such as the theory of gearing, ballistics and ship-construction.

924a. Grigor'ian, A.T., and L.S. Polak, "Die Grundideen der Mechanik von Heinrich Hertz." *NTM* (1964): Beiheft, 89-101.

An exposition of some of the leading ideas of Hertz' *Prinzipien der Mechanik* (1894).

925. Kargon, Robert, "William Rowan Hamilton and Boscovichean Atomism." *J. Hist. Ideas* 26 (1965): 137-140.

Argues that Hamilton's "first major effort to establish general mathematical laws of dynamics" in 1834 was based on a definite physical model of the Boscovichean atom, in keeping with Hamilton's early antipathy to a material substratum.

926. Mayr, Otto, "Maxwell and the Origins of Cybernetics." *Isis* 62 (1971): 425-444.

Analyzes a little known paper of Maxwell's, "On Governors" (1867/68), placing it in its intellectual context and arguing that both features of the paper, namely dynamic stability and differential gearing, were old interests of his. Further shows that the paper had little impact outside the field of rational mechanics, due to its highly mathematical and non-practical approach.

927. Mayr, Otto, "Victorian Physicists and Speed Regulation: An Encounter Between Science and Technology." *Notes Rec. Roy. Sci. Lond.* 26 (1971): 205-228.

Describes the work of six scientists on the theory and practical application of governors between 1840 and 1870, arguing that mutual scientific contact encouraged this burst of activity and further that there is a distinction between the earlier work by Airy, Siemens, Foucault and Thomson in which theory and practice are combined and the sharp separation of the two in the later approach of Maxwell and Gibbs.

928. Oravas, Gunhard ÂE., and Leslie McLean, "Historical Development of Energetical Principles in Elastomechanics. I: From Heraclitos to Maxwell; II: From Cotterill to Prange." *Applied Mechanics Reviews* 19 (1966): 647-58 and 919-33.

Brief summaries of the major contributions to the field, with no attempt to provide an historical synthesis. Good bibliography of primary sources.

929. Pogrebysskii, Iosif B., *Ot Lagranzha k Einshteinu: Klassicheskaya mekhanika XIX v.* (From Lagrange to Einstein: Classical Mechanics in the 19th Century.) Moscow: Nauka, 1966. 326 pp.

Describes the work of Lagrange, Carnot, Monge, Laplace, Poinsot, Hamilton, Jacobi, Ostrogradsky, Darboux, Beltrami and Lipschitz.

930. Poinsot, Louis, *La théorie générale de l'équilibre et du mouvement des systèmes.* Critical edition and commentaries by Patrice Bailhache. Paris: J. Vrin, 1975. 220 + 100 pp.

Concerned above all with the principle of virtual velocities, giving a comparative edition of various versions of Poinsot's 1806 paper on the subject and discussing various other attempts at proving the principle as well.

931. Polak, L.S., "The Development of the Principles of the Dynamics of System in the 19th and 20th Centuries." *Actes du VIIIe Congrès International d'Histoire des Sciences, Florence-Milan, 1956* (Florence, 1958), vol. 1, pp. 174-177.

Briefly traces the major developments in the analytical dynamics of generalized spaces, used extensively in solving physical problems involving electromagnetism, heat, statistical mechanics, etc.

932. Smith, G.C., "Matthew O'Brien's Anticipation of Vectorial Mathematics." *Hist. Math.* 9 (1982): 172-190.

Describes O'Brien's attempt in the 1840s to develop a vector algebra. Argues that his papers from this period contained most of the concepts requisite for this, but that not only did he fail to convince his contemporaries of the value of his ideas, he even subsequently discarded many of them himself.

933. Stephenson, Reginald J., "Development of Vector Analysis from Quaternions." *Amer. J. Phys.* 34 (1965): 194-201.

Describes the initial use and application of quaternions by Hamilton, Maxwell, and others, and the simplifications Gibbs introduced through his development of vectors.

933a. Wheeler, Lynde P., et al., eds., *The Early Work of Willard Gibbs in Applied Mechanics*. New York: Schuman, 1947. vii + 78 pp.

Comprises the text of Gibbs' previously unpublished Ph.D. thesis and accounts of his mechanical inventions.

(f) Light

934. Banet, Leo, "Evolution of the Balmer Series." *Amer. J. Phys.* 34 (1965): 496-503.

Suggests a background for the mathematician Balmer's derivation of his formula, namely the work of the physicist Hagenbach and the spectroscopist Kayser. Reconstructs and expands Balmer's original paper, suggesting that it derived from his work on geometry.

935.   Banet, Leo, "Balmer's Manuscripts and the Construction
       of His Series." *Amer. J. Phys.* 38 (1970): 821-828.

       By drawing on unpublished manuscripts, reconstructs
       the path by which Balmer derived his series.

936.   Buchwald, Jed Z., "Optics and the Theory of the Puncti-
       form Ether." *Arch. Hist. Exact Sci.* 21 (1979-80):
       245-278.

       Examines the development of the molecular aether
       model in optics, which started with Fresnel's work but
       was soon replaced by Cauchy's derivation of a three-
       dimensional lattice matrix model for the wave equation.
       At first accepted by the British in the 1830s, this was
       all but dropped due to specific problems and general
       opposition to molecular models. Cauchy, however, then
       (1850) devised a working theory of periodic lattice
       structures justifiable by a punctiform aether model.

*   Cantor, G.N., "The Changing Role of Young's Ether."

       Cited herein as item 760.

*   Cantor, G.N., "Henry Brougham and the Scottish Methodo-
       logical Tradition."

       Cited herein as item 761.

*   Cantor, G.N., "The Reception of the Wave Theory of
       Light in Britain: A Case Study Illustrating the Role
       of Methodology in Scientific Debate."

       Cited herein as item 762.

*   Cawood, John A., "The Scientific Work of D.F.J. Arago,
       1786-1853."

       Cited herein as item 714.

*   Chappert, A., "Deux lettres autographes d'Augustin
       Fresnel."

       Cited herein as item 716.

*   Chappert, A., *Etienne Louis Malus (1775-1812) et la
       théorie corpusculaire de la lumière.*

       Cited herein as item 717.

*   Chappert, A., "Lettres nouvelles de la correspondance
       de Fresnel."

       Cited herein as item 718.

* Chappert, A., "Poisson et les problèmes de l'optique: la controverse avec Fresnel."

  Cited herein in item 742.

937. Dingle, Herbert, "A Hundred Years of Spectroscopy." *Brit. J. Hist. Sci.* 1 (1962-63): 199-216.

  A brief, roughly chronological account of advances in knowledge and instrumentation in spectroscopy through the nineteenth century.

* Frankel, Eugene, "The Search for a Corpuscular Theory of Double Refraction: Malus, Laplace and the Prize Competition of 1808."

  Cited herein as item 726.

* Frankel, Eugene, "Corpuscular Optics and the Wave Theory of Light: The Science and Politics of a Revolution in Physics."

  Cited herein as item 727.

938. Gerlach, W., "Joseph Fraunhofer und seine Stellung in der Geschichte der Optik." *Optik* 20 (1963): 279-292.

  A brief survey, without documentation, of Fraunhofer's optical work.

939. Hargreave, David, "Thomas Young's Theory of Colour Vision: Its Roots, Development, and Acceptance by the British Community." Ph.D. dissertation, University of Wisconsin, 1973.

  Argues that Young's theory of colour vision developed as an adjunct to his theory of light, that it was evolved as part of a tradition going back to Descartes, and that it was not generally accepted until the 1850s after it had been modified by both Maxwell and Helmholtz.

940. Jaffe, Bernard, *Michelson and the Speed of Light*. Garden City, N.Y.: Doubleday, Anchor Books, 1960. (Science Study Series.) 197 pp.

  A readable account of Michelson's scientific life, with an emphasis on his aether-drift experiments. Reviewed by R. Siegfried, *Isis* 53 (1962): 426-428.

941. Kangro, Hans, "Kirchhoff und die spektralanalytische Forschung." *Untersuchungen über das Sonnenspectrum*

*und die Spectren der chemischen Elemente, und weitere
ergänzende Arbeiten aus den Jahren 1859-1862* by G.R.
Kirchhoff, ed. Hans Kangro. Osnabrück: Otto Zeller,
1972. Separately paginated, pp. 1-54.

Provides a good background history of spectral analysis and Kirchhoff's role in changing the discipline.
Includes a good primary and secondary bibliography.
The first part of the book consists of facsimiles of
eight of Kirchhoff's papers.

942. Kayser, Heinrich, "Geschichte der Spectroscopie." *Handbuch der Spectroscopie*. 8 vols. Leipzig: S. Hirzel
Verlag, 1900-32. Vol. 1, pp. 3-128.

A comprehensive account covering the whole of the 19th
century.

* Latchford, K.A., "Thomas Young's Work on Optics."

Cited herein as item 769a.

* Leitner, Alfred, "The Life and Work of Joseph Fraunhofer (1787-1826)."

Cited herein as item 73.

943. McGucken, William, *Nineteenth-Century Spectroscopy:
Development of the Understanding of Spectra, 1802-
1897*. Baltimore/London: Johns Hopkins University
Press, 1970. xii + 233 pp.

The bulk of this monograph is devoted to an analysis
of the writings of Bunsen and Kirchhoff and later
workers from 1850 onwards. The approach is purely
internalist. McGucken deals chiefly with flame rather
than solar spectra, the influence that spectroscopy
had on theories of matter, and attempts at its mathematization; he does not discuss instrument design and
development. Reviewed by A.J. Meadows, *Ann. Sci.* 27
(1972): 301-302 and J. Brookes Spencer, *Isis* 65 (1974):
284-285.

944. MacLean, J., "On Harmonic Ratios in Spectra." *Ann.
Sci.* 28 (1972): 121-137.

Discusses the evolution and development of the concept
of harmonic ratios between frequencies in the line
spectra of molecules, and the eventual invalidation of
the idea. Conjectures how the concept might later have
aided Balmer in formulating the "correct result."

945.   McRae, Robert J., "The Origin of the Conception of the
       Continuous Spectrum of Heat and Light." Ph.D. dis-
       sertation, University of Wisconsin, 1969.

       Discusses the work of Pictet, William Herschel, Leslie,
       Rumford, Prévost, Ritter, Seebeck, De la Roche, Powell,
       Forbes, Draper, Melloni and others in the period 1775-
       1850, examining both the observational and the conceptual
       disputes that frequently occurred as ideas concerning
       the relationship between radiant light and heat evolved.

*   Maier, Clifford L., *The Role of Spectroscopy in the
       Acceptance of the Internally Structured Atom, 1860-
       1920.*

       Cited herein as item 815.

*   Morse, Edgar William, "Natural Philosophy, Hypotheses,
       and Impiety: Sir David Brewster Confronts the Undu-
       latory Theory of Light."

       Cited herein as item 772.

*   Newburgh, Ronald, "Fresnel Drag and the Principle of
       Relativity."

       Cited herein as item 900.

*   Rohr, Moritz von, *Joseph Fraunhofers Leben, Leistungen
       und Wirksamkeit.*

       Cited herein as item 85.

946.   Ronchi, Vasco, "Schopenhauer con Goethe e contro Goethe
       in tema di colore." *Physis* 1 (1959): 279-293.

       Argues for Goethe's importance in opposing the attri-
       bution of colour to radiation alone and in calling for
       a physiologically based account of colour instead, and
       describes Schopenhauer's theory of colour as retinal
       activity.

*   Rosmorduc, Jean, "Ampère et l'optique: une intervention
       dans le débat sur la transversalité de la vibration
       lumineuse."

       Cited herein as item 745.

947.   Runge, Iris, *Carl Runge und sein wissenschaftliches
       Werk.* Göttingen: Vandenhoek und Ruprecht, 1949.
       214 pp. (Abhandlungen der Akademie der Wissenschaften
       zu Göttingen, Math.-Phys. Kl., 3rd series, no. 23).

The fullest and most authoritative source for Runge's
life, written by his daughter. Deals with Runge's extensive work on spectroscopy and applied mathematics
during the period 1887-1904 (his best known work being
on the Zeeman effect, 1900-1902). Runge's careful
methodical researches are contrasted with the bolder
and more productive work of Rydberg.

*   Schaffner, Kenneth L., ed., *Nineteenth-Century Aether
        Theories*.

        Cited herein as item 833.

948.   Shankland, R.S., "Michelson-Morley Experiment." *Amer.
        J. Phys.* 32 (1964): 16-35.

        Traces Michelson's professional career and the develop-
        ment of his interferometer, 1879-1887, before describing
        the Michelson-Morley experiment itself. Argues that
        the experiment was prompted by a letter from Maxwell to
        an associate of Michelson's, David Peck Todd, and was
        continued with the encouragement of Helmholtz, Kelvin
        and Rayleigh.

949.   Shankland, R.S., "Rayleigh and Michelson." *Isis* 58
        (1967): 86-88.

        Draws attention to the friendship between Rayleigh
        and Michelson stemming from Michelson's early optical
        work, and the support and recognition that the estab-
        lished Rayleigh gave to the then unknown American.

949a.   Sherman, Paul D., *Colour Vision in the Nineteenth
        Century: The Young-Helmholtz-Maxwell Theory*. Bristol:
        Adam Hilger Ltd., 1981. xiv + 233 pp.

        A detailed account of the development of the three-
        color theory of color vision during the nineteenth
        century. In addition to the work of Young, Helmholtz
        and Maxwell, the contributions of Brewster and Grassmann
        are also discussed. Maxwell's work is seen as supremely
        important: his "experimental methods did nothing less
        than transform a qualitative science into a quantitative
        one," and thus constitute the starting point of the
        subject in its modern form.

*   Silliman, Robert H., "Fresnel and the Emergence of
        Physics as a Discipline."

        Cited herein as item 747.

950. Sutton, M.A., "Sir John Herschel and the Development of Spectroscopy in Britain." *Brit. J. Hist. Sci.* 7 (1974): 42-60.

Looks at the sporadic experimental and theoretical work done on spectra in Britain between 1817 and 1860. Argues that in principle spectroscopy had importance for theories of both matter and light, especially in relation to the "law of continuity," but that it was largely ignored until taken up by the chemists.

951. Sutton, M.A., "Spectroscopy and the Chemists: A Neglected Opportunity?" *Ambix* 23 (1976): 16-26.

Uses the failure of nineteenth-century chemists to exploit much sooner than they did the possibilities of spectroscopic analysis to support some interesting comments on the problems posed by interdisciplinary boundaries between the different sciences—in this case the boundary between chemistry and physics.

* Swenson, Loyd S., Jr., *The Etherial Aether: A History of the Michelson-Morley-Miller Aether-Drift Experiments, 1880-1930.*

Cited herein as item 903.

* Thiele, Joachim, "Zur Wirkungsgeschichte des Dopplerprinzips im Neunzehnten Jahrhundert."

Cited herein as item 842.

* Van Broeckhoven, R.L.J.M., "The Growth of Thomas Young's Ideas on Interference."

Cited herein as item 776.

* Wilson, David B., "The Reception of the Wave Theory of Light by Cambridge Physicists (1820-50): A Case Study in the Nineteenth-Century Mechanical Philosophy."

Cited herein as item 845.

952. Wilson, D.B., "George Gabriel Stokes on Stellar Aberration and the Luminiferous Ether." *Brit. J. Hist. Sci.* 6 (1972-73): 57-72.

Describes Stokes' papers on aberration (1845-1848) and the dispute they engendered with James Challis, arguing that Stokes rejected Fresnel's theory because of his earlier hydrodynamical research (which had made

him "acutely conscious of the results of mutual fric-
tion" in solid-fluid interactions) and his view of the
aether as an elastic solid.

*   Wood, Alexander, *Thomas Young, Natural Philosopher,
        1773-1829.*

        Cited herein as item 107.

*   Worrall, John, "Thomas Young and the 'Refutation' of
        Newtonian Optics: A Case Study of the Interaction of
        Philosophy of Science and History of Science."

        Cited herein as item 779.

                            (g) Heat

(i) General, Nature of Heat

953.   Bachelard, Gaston, *Etude sur l'évolution d'un problème
        de physique: la propagation thermique dans les solides.*
        Paris: 1927. Reprinted, Paris: J. Vrin, 1973. v +
        183 pp.

        Chiefly concerned with 19th-century theories. Pri-
        marily concerned with mathematical theories of heat con-
        duction rather than the physical models assumed, dealing
        particularly with the work of Biot, Fourier, Comte,
        Poisson, Duhamel, Lamé and Boussinesq.

*   Bellone, Enrico, "Il significato metodologico dell'
        eliminazione dei modelli del calorico promossa da
        Joseph Fourier."

        Cited herein as item 712.

954.   Brush, Stephen G., "The Wave Theory of Heat: A Forgotten
        Stage in the Transition from the Caloric Theory to
        Thermodynamics." *Brit. J. Hist. Sci.* 5 (1970-71):
        145-167. Reprinted in *The Kind of Motion We Call
        Heat* by Stephen G. Brush. Amsterdam: North-Holland,
        1976. Vol. 2, pp. 303-325.

        Shows that between 1830 and 1850 a large number of
        physicists espoused a wave theory of heat according to
        which heat consisted of vibrations in the caloric or
        aether. The theory rested on the perceived analogy
        between the properties of radiant heat and those of

light, and on the wave theory of light. Argues that it may in fact have contributed to the discovery of conservation of energy.

955. Burchfield, Joe D., *Lord Kelvin and the Age of the Earth.* London/New York: Science History Publications, 1975. xii + 260 pp.

Describes the debate on the age of the Earth in the latter half of the 19th century, taking as the central point of interest Kelvin's thermodynamically based value and the responses it aroused. Reviewed by H.I. Sharlin, *Ann. Sci.* 33 (1976): 485–488, and J.Z. Buchwald, *Isis* 67 (1976): 492–494.

956. Cornell, E.S., "The Radiant Heat Spectrum from Herschel to Melloni. II: The Work of Melloni and His Contemporaries." *Ann. Sci.* 3 (1938): 402–416.

Concentrates on Melloni's work based on his development of the thermocouple, and shows how evidence for the identity of heat and light rays gradually accumulated in the period 1831–1853. For the earlier part of the story, see item 607.

*   Finn, Bernard S., "Laplace and the Speed of Sound."

Cited herein as item 723.

957. Fox, Robert, "The Background to the Discovery of Dulong and Petit's Law." *Brit. J. Hist. Sci.* 4 (1968–69): 1–22.

Argues that Dulong, Petit and Berzelius played a significant part in the overthrow of the caloric theory by proposing a viable, alternative electrochemical model of heat.

*   Friedman, Robert Marc, "The Creation of a New Science: Joseph Fourier's Analytical Theory of Heat."

Cited herein as item 730.

*   Grattan-Guinness, I., "Joseph Fourier and the Revolution in Mathematical Physics."

Cited herein as item 734.

*   Grattan-Guinness, Ivor, in collaboration with J.R. Ravetz, *Joseph Fourier 1768-1830: A Survey of His Life and Work, based on a Critical Edition of His Monograph*

on the Propagation of Heat, Presented to the Institut
de France in 1807.

Cited herein as item 735.

* Herivel, John, *Joseph Fourier: The Man and the Physicist.*

Cited herein as item 62.

* Herivel, John, *Joseph Fourier face aux objections contre
sa théorie de la chaleur: lettres inédites 1808-1816.*

Cited herein as item 740.

958. James, Frank A.J.L., "Thermodynamics and Sources of
Solar Heat, 1846-1862." *Brit. J. Hist. Sci.* 15 (1982):
155-181.

Argues that the theoretical basis of William Thomson's
famous attack on the longevity that Darwin had proposed
for the Earth was not developed specifically as a response
to Darwin, but arose somewhat earlier out of a widespread
recognition that the newly established laws of thermodynamics raised serious questions about possible sources
of solar heat.

959. Kangro, Hans, "Ultrarotstrahlung bis zur Grenze elek-
trisch erzeugter Wellen: Das Lebenswerk von Heinrich
Rubens." *Ann. Sci.* 26 (1970): 235-259.

Describes Rubens' work from the 1890s to 1910 on heat
radiation, optics, and especially in linking electromagnetic theory to infra-red radiation.

960. Kangro, Hans, *Early History of Planck's Radiation Law.*
Translated by R.E.W. Maddison in collaboration with
the author. London: Taylor and Francis, 1976. xvii +
282 pp.

A revised translation of the author's *Vorgeschichte
des Planckschen Strahlungsgesetzes* (1970). Drawing on
manuscript sources, Kangro has produced a comprehensive
account of the intensive work between 1879 and 1900 on
the energy distribution law, from Stefan's T^4 law,
through Wien's displacement law to Planck's formula and
his explanation of it in terms of the quantum.

961. Kargon, R., "The Decline of the Caloric Theory of Heat:
A Case Study." *Centaurus* 10 (1964): 35-39.

Argues for the influence of Kant on the decline of
imponderable-fluid theories, via a case study of the
work of the Russian chemist A.N. Scherer.

*   Körber, Hans-Günther, "Aus der Entwicklung der Bimetall-
      thermometer zu Anfang des 19. Jahrhundert."

      Cited herein as item 886a.

*   Kuhn, Thomas S., "The Caloric Theory of Adiabatic Com-
      pression."

      Cited herein as item 1019.

*   McRae, Robert J., "The Origin of the Conception of the
      Continuous Spectrum of Heat and Light."

      Cited herein as item 945.

962. Olson, Richard G., "A Note on Leslie's Cube in the Study
      of Radiant Heat." *Ann. Sci.* 25 (1969): 203-208.

      Describes the uses of the cube, from 1804 to the
      1880s, first by Leslie to derive the sine law of radia-
      tion from surfaces and the relative emissivity of dif-
      ferent surfaces, and later by others in treating the
      transmission of heat through various media and the
      proportionality of emissivity and absorptivity.

963. Olson, Richard G., "Count Rumford, Sir John Leslie, and
      the Study of the Nature and Propagation of Heat at
      the Beginning of the Nineteenth Century." *Ann. Sci.*
      26 (1970): 273-304.

      Using Leslie's and Rumford's theories as exemplars of,
      respectively, the material and dynamical models of heat,
      Olson argues that prior to more sophisticated work *c.*
      1850, the material theories were as powerful a research
      tool, and would account for phenomena (i.e. propagation
      of heat, specific and latent heat, expansion effects)
      at least as well as the dynamical theory, with the ex-
      ception, to a certain extent, of friction.

963a. Roller, Duane H.D., "Thilorier and the First Solidifica-
      tion of a 'Permanent' Gas (1835)." *Isis* 43 (1952):
      109-113.

      Includes a brief account of the circumstances of
      Thilorier's experiment, some biographical details, a
      facsimile of the original account of the work and an
      English translation of this.

963b. Schöpf, Hans-Georg, *Von Kirchhoff bis Planck: Theorie der Wärmestrahlung in historisch-kritischer Darstellung.* Braunschweig: Vieweg, 1978. 199 pp.

Not seen.

963c. Sebastiani, Fabio, "Le teorie caloricistiache di Laplace, Poisson, Sadi Carnot, Clapeyron e la teoria dei fenomeni termici nei gas formulata da Clausius nel 1850." *Physis* 23 (1981): 397-438.

Briefly reviews the caloric theories of Laplace, Poisson, Carnot and Clapeyron before analyzing Clausius' famous 1850 memoir on the motive power of heat to display "the osmosis between caloric and dynamical theories of thermal phenomena in gases."

964. Siegel, Daniel M., "Balfour Stewart and Gustav Robert Kirchhoff: Two Independent Approaches to 'Kirchhoff's Radiation Law.'" *Isis* 67 (1976): 565-600.

Describes and analyzes Stewart's and Kirchhoff's accounts of the Radiation Law (1858 and 1859)--showing how Stewart stressed experimental and pragmatic elements and Kirchhoff, rigorous theoretical demonstrations--and also the acrimonious 40-year debate which ensued concerning priority for an adequate account of the law. Shows that the debate, while fundamentally void of constructive scientific criticism, reflects parochial conceptions of the standards for judging scientific work, and was part of a more general debate over nationalistic scientific reputations.

(ii) Energy Conservation

965. Bluh, Otto, "The Value of Inspiration: A Study on Julius Robert Mayer and Josef Popper-Lynkeus." *Isis* 43 (1952): 211-220.

Publishes translations of letters from Mayer to Popper which show how much he desired to be given priority for the notion of conservation of energy. Includes an appendix giving a bibliography of the priority controversy involving Tyndall, Thomson and Tait, Joule, Colding etc. (1862-64).

966. Cantor, G.N., "William Robert Grove, the Correlation of Forces, and the Conservation of Energy." *Centaurus* 19 (1975): 273-290.

A discussion of Grove's philosophy of natural science
and an elucidation of his doctrine. Compares and contrasts this doctrine with the conservation of energy
principle.

967. Colding, Ludvig, *Ludvig Colding and the Conservation
of Energy Principle: Experimental and Philosophical
Contributions.* Trans. with an introduction and commentary by Per F. Dahl. New York/London: Johnson Reprint, 1972. xxxvi + 197 pp. (Sources of Science,
104).

Contains "essentially all" of Colding's papers (1850-
1864) concerned with the conservation of energy, built
around his principal notion that "forces" are not
destroyed but transformed. Dahl's introduction provides
a brief biography, a background for Colding's ideas and
an analysis of his work.

968. Dahl, Per F., "Ludvig Colding and the Conservation of
Energy." *Centaurus* 8 (1963): 174-188.

. Examines the role of Colding's somewhat neglected
work in the formulation of the energy principle. Oersted's work is also discussed.

969. Elkana, Yehuda, "The Conservation of Energy: A Case of
Simultaneous Discovery?" *Arch. Int. Hist. Sci.* 23
(1970): 31-60.

Argues that the different "discoverers" of the principle in fact were working on different problems and
came up with different solutions--Joule on the mechanical
equivalent of heat, Thomson on reconciling Joule and
Carnot, Rankine on whether to adopt a mechanical or a
dynamical theory of heat, and Helmholtz on a mathematical
conservation law--and it is only in hindsight that these
are seen as simultaneous discoveries of the same concept.

970. Elkana, Yehuda, "Helmholtz' 'Kraft': An Illustration of
Concepts in Flux." *Hist. Studs. Phys. Sci.* 2 (1970):
263-298.

Uses Helmholtz' development of the conservation of
energy principle as an exemplar of how theory and concept necessarily crystallize together. Also reviews
contemporary efforts in England and Germany to come
to terms with the concepts of force and energy.

971. Elkana, Yehuda, *The Discovery of the Conservation of
        Energy*. London: Hutchinson, 1974. x + 213 pp.

        Taking Helmholtz' work as the culmination of the story,
        looks at the confused emergence of the concept of
        energy (especially in relation to that of "force") out
        of separate traditions in mechanics, heat theory and
        physiology. Then attempts to fit the discovery into an
        institutional and philosophical background. Reviewed
        by H.J. Steffens, *Isis* 67 (1976): 137-139, and G.N.
        Cantor, *Brit. J. Hist. Sci.* 8 (1975): 87-88.

972. Forrester, J., "Chemistry and the Conservation of
        Energy: the Work of James Prescott Joule." *Studs.
        Hist. Phil. Sci.* 6 (1975): 273-313.

        Shows that Joule's idea of a mechanical equivalence
        of heat arose in 1844 from problems he had with an
        electrochemical theory of heat that gave ontological
        primacy to electricity. The significance of the dis-
        covery was not recognized until Joule's fortuitous
        presentation of a paper to the Physics rather than the
        Chemistry section of the BAAS meeting of 1847.

973. Gooding, David, "Metaphysics Versus Measurement: The
        Conversion and Conservation of Force in Faraday's
        Physics." *Ann. Sci.* 37 (1980): 1-29.

        Demonstrates the interlinking of Faraday's chemistry
        and physics with his theology, resulting in his rejecting
        the apparent materialism of the "new mechanical philoso-
        phy" and its concepts, and his adopting of a "wave
        model" for the conversion and transmission of electro-
        magnetic phenomena and heat.

974. Heimann, P.M., "Conversion of Forces and the Conserva-
        tion of Energy." *Centaurus* 18 (1974): 147-161.

        Argues *contra* Kuhn (item 979) that the notion of in-
        destructability and interconvertibility of forces such as
        heat, light and electricity was not specific to 1830-
        1850, but was an accepted part of British natural
        philosophy from at least the end of the 18th century.

975. Heimann, P.M., "Helmholtz and Kant: The Metaphysical
        Foundations of *Über die Erhaltung der Kraft*." *Studs.
        Hist. Phil. Sci.* 5 (1974): 205-238.

        Discusses Helmholtz' indebtedness to Kant's philosophi-
        cal ideas. Also deals with the roots of Helmholtz'
        memoirs on physiology.

976. Heimann, P.M., "Mayer's Concept of 'Force': the 'Axis'
of a New Science of Physics." *Hist. Studs. Phys.
Sci.* 7 (1976): 277-296.

Argues that Mayer defined forces in terms of both
their phenomenal effects--that forces were qualitatively
transformable but quantitatively constant--and their
ontological status, as phenomenal manifestations of a
single *Urkraft* or basic force distinct from matter in
being indestructible but not substantial. This led him
to conservation concepts but also to maintain that heat
was a manifestation of the *Urkraft* distinct from motion.
Points out the similarities but also the differences
between Mayer's conceptions and those of Leibniz, Kant
and the proponents of *Naturphilosophie*.

977. Hermann, Armin, "Die Entdeckung des Energie-Prinzips:
Wie der Arzt Julius Robert Mayer die Physiker belehrte."
Bild der Wissenschaft 4 (1978): 140-148.

A brief undocumented discussion of Mayer's claim to
priority as a discoverer of the law of energy conservation. Does, however, include some useful analysis of
the development of Mayer's ideas around 1841-42, including
his exchanges with Baur and Nörremberg.

978. Jones, G., "Joule's Early Researches." *Centaurus* 13
(1968): 198-219.

Provides various views of the path followed by Joule
from his early work to the energy principle. Argues
that Joule's interest in electromagnetic engines was
important.

979. Kuhn, Thomas S., "Energy Conservation as an Example of
Simultaneous Discovery." *Critical Problems in the
History of Science*, ed. Marshall Clagett. Madison:
University of Wisconsin Press, 1959, pp. 321-356.
Reprinted in *The Essential Tension: Selected Studies
in Scientific Tradition and Change* by Thomas S. Kuhn.
Chicago/London: University of Chicago Press, 1977,
pp. 66-104.

Proposes three common influences--the consciousness
of energy conversion processes, the technical discussion
of dynamical engines, and the unificatory ideal of
Naturphilosophie--to account, at least partially, for
the simultaneous discovery of energy conservation by
Mayer, Joule, Colding, Helmholtz and others during the
period 1842-1847. Also discusses the inherent difficulties

of the term "simultaneous discovery."

980.   Lindsay, Robert Bruce, *Julius Robert Mayer: Prophet of
         Energy*. Oxford/New York: Pergamon Press, 1973.
         (Selected Readings in Physics: Men of Physics.)
         vii + 238 pp.

         Provides translations of five of Mayer's papers, and
         also includes a brief biographical sketch and a commen-
         tary on his work. As implied by the title, Lindsay is
         rather Whiggish in his approach, and reads into Mayer's
         work a full, modern concept of energy which he did not
         hold. This is implicit, most notably, in the consistent
         rendering of "Kraft" as "energy."

981.   Lloyd, J.T., "Background to the Joule-Mayer Controversy."
         Notes Rec. Roy. Soc. Lond. 25 (1970): 211-225.

         Provides some personal background to the dispute
         (1862-63) between Tait and Tyndall, on behalf of Joule
         and Mayer respectively, concerning priority for the
         principle of conservation of energy, by quoting exten-
         sively from previously unpublished letters of Joule
         and Tait to Thomson.

982.   Mendoza, E., and D.S.L. Cardwell, "On a Suggestion Con-
         cerning the Work of J.P. Joule." *Brit. J. Hist. Sci.*
         14 (1981): 177-180.

         Critical comments on H.J. Steffens' *James Prescott
         Joule and the Concept of Energy* (New York, 1979) (item
         986), based on fresh evidence.

982a. Merleau-Ponty, Jacques, "La découverte des principes
         de l'energie: l'itinéraire de Joule." *Rev. Hist. Sci.*
         32 (1979): 315-331.

         Analyzes the series of papers published by Joule in
         the period 1840-1843 leading up to his notion of the
         equivalence of heat and mechanical energy and his suc-
         cessful measurement of the mechanical equivalent, J.

983.   Rossi, Arcangelo, "L'esperimento di Joule sull'equivalente
         meccanico del calore e il secondo principio della
         termodinamica." *Physis* 19 (1977): 337-353.

         Argues that although by virtue of its greater explana-
         tory power Joule's dynamical theory was accepted, it
         was only able to be connected with Carnot's Principle
         with Clausius' reformulation of the Principle.

984. Smith, Crosbie W., "Faraday as a Referee of Joule's
     Royal Society Paper 'On the Mechanical Equivalent of
     Heat.'" *Isis* 67 (1976): 444-449.

     Argues that, due to Faraday's growing commitment to
     the principle of conservation of force, though he agreed
     with Joule that there was a proportionality between
     mechanical force and heat, he disagreed absolutely
     with Joule's theory of the interconvertibility of the
     two.

985. Smith, Crosbie W., "A New Chart for British Natural
     Philosophy: The Development of Energy Physics in the
     Nineteenth Century." *Hist. Sci.* 16 (1978): 231-279.

     Argues that "William Thomson ... was the central
     figure in laying the foundations of energy physics in
     Britain during the period 1850-75," and that "simple
     internalist or externalist historiographical categories
     are inadequate for an understanding of the ways in
     which energy physics was important to nineteenth-century
     British physical scientists." The study is intended to
     exemplify a more satisfactory alternative.

986. Steffens, Henry John, *James Prescott Joule and the Con-
     cept of Energy.* New York: Science History Publica-
     tions/Folkestone: Dawson, 1979. x + 173 pp.

     An examination of Joule's work derived from the author's
     doctoral thesis. Includes discussion of Joule's early
     investigations of a theory of heat based on a connection
     between heat and electricity, an account of the Joule-
     Mayer controversy, an analysis of the mechanical equivalence
     of heat experiments and how they were tied to the
     theoretical work of Thomson, Rankine and Clausius, and
     finally a short section on the significance of Joule's
     work for the evolution of the concept of energy. Re-
     viewed by Crosbie Smith, *Ambix* 27 (1980): 142, and J.D.
     Burchfield, *Isis* 71 (1980): 183. Cf. also item 982 above.

987. Wise, M. Norton, "William Thomson's Mathematical Route
     to Energy Conservation: A Case Study of the Role of
     Mathematics in Concept Formation." *Hist. Studs. Phys.
     Sci.* 10 (1979): 49-84.

     Gives the history of the mathematical physics of
     energy conservation, and argues that Thomson's physical
     conceptualization of mathematical expressions in his
     natural philosophy was instrumental in his creation of
     new physical ideas.

(iii) Thermodynamics

988.   Barnett, Martin K., "Sadi Carnot and the Second Law of
       Thermodynamics." *Osiris* 13 (1958): 327-357.

       A useful description of the Carnot cycle and the prob-
       lems Carnot faced in his work.

989.   Birembaut, A., "A propos des notices biographiques sur
       Sadi Carnot: quelques documents inédits." *Rev. Hist.
       Sci.* 27 (1974): 355-369.

       Using previously unpublished documents, corrects and
       augments the standard biographical accounts of Sadi
       Carnot, and shows the role of his brother Hippolyte
       and nephew Adolphe in the publication of his work.

990.   Brunold, Charles, *L'entropie: son role dans le développe-
       ment historique de la thermodynamique.* Paris: Masson,
       1930. 221 pp.

       A general history of the evolution and development of
       the concept of entropy, concentrating mainly on the work
       of Carnot, Clausius, and Rankine.

991.   Cardwell, D.S.L., *From Watt to Clausius: The Rise of
       Thermodynamics in the Early Industrial Age.* London:
       Heinemann, 1971. xvi + 336 pp.

       A detailed study of the rise of thermodynamics which
       focuses upon the interrelations between scientific
       developments and the advancement of steam power tech-
       nology. Reviewed by O. Mayr, *Isis* 63 (1972): 451-452,
       and J. Payen, *Brit. J. Hist. Sci.* 7 (1974): 171-175.

992.   Cardwell, D.S.L., "Science and the Steam Engine, 1790-
       1825." *Science and Society, 1600-1900*, ed. Peter
       Mathias. Cambridge: Cambridge University Press,
       1972, pp. 81-96.

       Summarizes various ideas set out in the author's book,
       From Watt to Clausius (item 991), surveying the develop-
       ment of steam engine technology in the period stated
       and arguing that the rise of thermodynamics, and Carnot's
       work in particular, "is unintelligible without reference
       to the technological developments that preceded and
       accompanied it."

993.   Cardwell, D.S.L., and Richard L. Hills, "Thermodynamics
       and Practical Engineering in the Nineteenth Century."
       Hist. Tech. 1 (1976): 1-20.

A general survey that focuses on developments in
steam engine technology after 1850.

994. Carnot, Sadi, *Réflexions sur la puissance motrice du
feu*, édition critique avec introduction et commentaire,
augmentée de documents d'archives et de divers manu-
scrits de Carnot, par Robert Fox. Paris: Vrin, 1978.
371 pp.

A critical edition of the *Réflexions* together with a
substantial introduction, exemplary scholarly apparatus,
and the texts of Carnot's various unpublished manuscripts
not only on thermodynamics but also on political economy,
religion and morality. Essay review by Eric Mendoza,
Brit. J. Hist. Sci. 14 (1981): 75-78.

995. [Carnot, Sadi], *Reflections on the Motive Power of Fire,
by Sadi Carnot, and Other Papers on the Second Law of
Thermodynamics by E. Clapeyron and R. Clausius*, ed.
with an introduction by E. Mendoza. New York: Dover,
1960. xxii + 152 pp.

Contains, in addition to an English translation of
Carnot's *Réflexions*, important selections from his
later, unpublished manuscripts as well as English
translations of Clapeyron's and Clausius' principal
memoirs on the second law.

996. Cimbleris, Borisas, "Reflections on the Motive Power of
a Mind." *Physis* 9 (1967): 393-420.

A discussion of Sadi Carnot's life and work.

997. Daub, Edward E., "Atomism and Thermodynamics." *Isis* 58
(1967): 293-303.

Argues that Clausius and Rankine both thought that
the heat added to a body had two distinct internal energy
functions instead of the one associated with modern classical thermodynamics. Briefly examines their two different
atomic models associated with these energy functions
and how each was used to derive the second law.

998. Daub, Edward E., "Entropy and Dissipation." *Hist.
Studs. Phys. Sci.* 2 (1970): 321-354.

Describes the somewhat confused process by which dissipation and entropy became bound together through the
work of Tait, Thomson, Clausius and finally Gibbs and
Maxwell, and also, as background to the confusion, the
violent priority controversy between Tait and Clausius.

999.   Daub, Edward E., "The Regenerator Principle in the
       Stirling and Ericsson Hot Air Engines." *Brit. J.
       Hist. Sci.* 7 (1974): 259-277.

       Describes the principle of the regenerator--to avoid
       the heat loss of a steam engine condenser--and the
       accepted negative theoretical appraisal of it by Hirn
       and Zeuner, despite the favorable (and more accurate)
       analysis of Rankine. The ultimate failure of the device
       was in fact due to inadequate metal technology.

1000. Daub, Edward E., "Sources for Clausius' Entropy Concept:
       Reech and Rankine." *Proceedings, 15th International
       Congress of the History of Science, Edinburgh, 10-19
       August, 1977*, ed. Eric G. Forbes. Edinburgh: Edin-
       burgh University Press, 1978, pp. 342-358.

       Makes a convincing case for the impact of the work of
       Reech and Rankine on Clausius' conceptual development.
       One important part of the conclusion is vitiated by an
       error of dating; footnote 42 should read 1852, not
       1862.

*   Donnan, F.G., and A. Haas, eds., *A Commentary on the
       Scientific Writings of J. Willard Gibbs*.

       Cited herein as item 795.

*   Fétizon, M., "La thermodynamique de Galilée à Clausius:
       influence de Gay-Lussac sur Carnot."

       Cited herein in item 721.

1001. Fox, Robert, "Watt's Expansive Principle in the Work
       of Sadi Carnot and Nicolas Clément." *Notes Rec. Roy.
       Soc. Lond.* 24 (1969): 233-253.

       Discusses the work of Clément and Desormes (1819)
       explaining Watt's principle as utilized in the recently
       introduced high-pressure steam engines. Further, by
       comparison of published and unpublished sources, argues
       that Carnot was strongly influenced by their unique
       theory of an adiabatic rather than isothermal final
       phase of expansion, and by their calculation of its
       mechanical power.

1002. Gabbey, W.A., and John Herivel, "Un manuscrit inédit de
       Sadi Carnot." *Rev. Hist. Sci.* 19 (1966): 151-166.

       Publication, with commentary, of a manuscript con-
       cerning a formula for "la puissance motrice de la Vapeur

d'Eau." The authors argue that this was written after 1816, but prior to the *Réflexions*.

1003. Garber, Elizabeth, "James Clerk Maxwell and Thermo-dynamics." *Amer. J. Phys.* 37 (1969): 146-155.

Shows that Maxwell was first attracted to thermo-dynamics research by Gibbs' work of 1873 (at least partly on account of its geometrical method), that he developed his own ideas on heterogeneous substances (which remained unpublished), and that he did much to publicize Gibbs' work.

1004. Hatton, A.P., and L. Rosenfeld, "An Analysis of Joule's Experiments on the Expansion of Air (including Original Results not Previously Published)." *Centaurus* 4 (1955-56): 310-318.

The authors provide a theoretical analysis verifying Joule's experiment establishing that in an isothermic achievement of equilibrium between high and low pressure containers of gas, the heat gained by the former equals that lost by the latter.

1005. Hiebert, Erwin N., *The Conception of Thermodynamics in the Scientific Thought of Mach and Planck.* Freiburg i. Br.: Ernst-Mach-Institut der Fraunhofer-Gesell-schaft zur Forderung der angewandten Forschung, 1968. 106 pp. (Ernst-Mach-Institut Bericht Nr. 5/68).

A discussion of the various thermodynamical publica-tions of Mach and Planck, organized around the theme of the growing difference of opinion between the two over the nature of physical theory, as especially evident in their respective inclinations towards con-ventionalist and realist interpretations of the first and second laws of thermodynamics.

1006. Hoyer, Ulrich, "Über den Zusammenhang der Carnotschen Theorie mit der Thermodynamik." *Arch. Hist. Exact Sci.* 13 (1974): 359-375.

A discussion of Carnot's 1824 calculations on the motive power of heat, and of the relationship between those calculations and the estimate of the mechanical equivalent of heat set out in his posthumous papers.

1007. Hoyer, Ulrich, "How did Carnot Calculate the Mechanical Equivalent of Heat?" *Centaurus* 19 (1975): 207-219.

Discusses previous conjectures on how Carnot's value
(not discovered until 46 years after his death) was
achieved, criticizes them, and proposes his own,
similar to that of Mach (1896).

1008.   Hoyer, Ulrich, "Das Verhältnis der Carnotschen Theorie
        zur Klassichen Thermodynamik." *Arch. Hist. Exact
        Sci.* 15 (1975-76): 149-197.

        This paper continues the exploration of the relation-
        ship between Carnot's theory and classical thermody-
        namics, along the lines developed by P. Lervig in his
        paper "On the Structure of Carnot's Theory of Heat"
        (item 1022). Explores a number of specific problems,
        comparing the treatments provided by the two theories.

1009.   Hoyer, Ulrich, "Considerations on Carnot's Mechanical
        Equivalent of Heat." *Proceedings, 15th International
        Congress for the History of Science, Edinburgh, 10-19
        August, 1977*, ed. Eric G. Forbes. Edinburgh: Edin-
        burgh University Press, 1977, pp. 359-367.

        A brief discussion of a possible method used by
        Carnot to calculate the value of the mechanical equiva-
        lent of heat which appears in his posthumous notes.

1010.   Hoyer, Ulrich, "Über Waterstons mechanisches Wärm-
        eäquivalent." *Arch. Hist. Exact Sci.* 19 (1978):
        371-381.

        Examines Waterston's calculations, comparing them
        with those of other physicists.

1011.   Hutchison, Keith R., "Der Ursprung der Entropiefunktion
        bei Rankine und Clausius." *Ann. Sci.* 30 (1973):
        341-364.

        Shows how the quantification of the dissipation of
        energy was indirectly evolved through the limited en-
        tropy function of Rankine. The applicability of this
        was expanded (rather obscurely) by Clausius, leading
        to his discovery of the law of entropy increase which
        actually expresses dissipation.

1012.   Hutchison, Keith R., "Mayer's Hypothesis: A Study of
        the Early Years of Thermodynamics." *Centaurus* 20
        (1976): 279-304.

        A discussion of the role of "Mayer's Hypothesis"
        (equivalent to supposing that the internal energy of

a perfect gas is a function only of absolute temperature) in the thermodynamics of the years 1847-1855.
Gives the background to the Joule-Thomson experiment
which definitively sorted out the problems associated
with Mayer's hypothesis.

1013. Hutchison, Keith R., "Rankine, Atomic Vortices, and the
       Entropy Function." *Arch. Int. Hist. Sci.* 31 (1981):
       72-134.

       A detailed analysis of the model of the structure of
       matter, the vortex atom, which led Rankine to the dis-
       covery of the entropy function in the early 1850s.
       Includes a close study of Rankine's attempts to free
       his results from dependence on this model, and examines
       the differences between the resulting theory and con-
       ventional thermodynamics. Aimed at the specialist
       historian of mathematical physics. Bibliography
       minimal.

1014. Hutchison, Keith R., "W.J.M. Rankine and the Rise of
       Thermodynamics." *Brit. J. Hist. Sci.* 14 (1981):
       1-26.

       A general survey of Rankine's unusual theories of
       heat and their place in the history of thermodynamics.
       Detailed argument is excluded, but the footnotes func-
       tion as a thorough annotated bibliography. Argues
       that despite Rankine's success in applying theoretical
       thermodynamics to heat-engine technology, he had
       little impact on the theory as it finally emerged,
       except for his explicit introduction of the entropy
       function in 1854. Although most of Rankine's ideas
       about heat are now forgotten, his contemporaries saw
       him as one of the founders of the subject.

1015. Keller, K., "Gustav Adolf Hirn, sein Leben und sein
       Werk." *Beiträge zur Geschichte der Technik* 3 (1911):
       20-60.

       Includes considerable discussion of Hirn's scien-
       tific work, which was chiefly in thermodynamics.

1016. Kerker, Milton, "Sadi Carnot and the Steam Engine
       Engineers." *Isis* 51 (1960): 257-270.

       Argues that Carnot's *Réflexions* (1824) was tech-
       nically relevant and well within the competence of
       the engineers of his day, and furthermore was studied
       by at least one leading engineer at the time. Therefore,

neither its anonymity nor its supposed complexity are
acceptable explanations for its total lack of influence
on contemporaries.

1017. Klein, Martin J., "Gibbs on Clausius." *Hist. Studs.*
       Phys. Sci. 1 (1969): 127-149.

       Discusses Gibbs' obituary of Clausius (1889), which
       provides insights into Gibbs' attitude toward various
       concepts of thermodynamics, especially the contentious
       issues of entropy and disgregation.

1018. Klein, Martin J., "The Early Papers of J. Willard
       Gibbs: A Transformation of Thermodynamics." *Proceed-*
       ings, 15th International Congress for the History of
       Science, Edinburgh, 10-19 August, 1977, ed. Eric G.
       Forbes. Edinburgh: Edinburgh University Press, 1978,
       pp. 330-341.

       Suggests (i) that Gibbs' initial pair of papers on
       thermodynamics mark an important stage in the accep-
       tance of entropy as a fundamental physical quantity,
       and (ii) that Gibbs' interest may well have been
       stimulated by Andrews' discovery of critical phenomena.

1019. Kuhn, Thomas S., "The Caloric Theory of Adiabatic Com-
       pression." *Isis* 49 (1958): 132-140.

       Argues that while adiabatic phenomena were relatively
       well known from the mid-18th century, research was
       diffuse and largely ignored until Laplace comprehensively
       linked it with his work on the theory of sound under
       the aegis of the caloric theory, with which it seemed
       to be in good agreement.

1020. Kuhn, Thomas S., "Engineering Precedent for the Work of
       Sadi Carnot." *Arch. Int. Hist. Sci.* 13 (1960): 251-
       255.

       Argues that Carnot's ideas evolved from the engineering
       tradition of Parent, Smeaton, Coulomb, Lazare Carnot and
       Hachette, not from that of theoretical physics.

1021. Kuhn, Thomas S., "Sadi Carnot and the Cagnard Engine."
       Isis 52 (1961): 567-579.

       Argues that the Cagnard heated air engine with its
       simple transference of pure heat into mechanical work
       may have influenced Carnot's conception of his ideal
       gas engine. It is similar in concept, it "could have

led him to several of the less precedented conclusions"
of his memoir, and he "almost certainly" knew of it.
(See also E. Mendoza's article, item 1025.)

1022. Lervig, P., "On the Structure of Carnot's Theory of
Heat." *Arch. Hist. Exact Sci.* 9 (1972-73): 228-239.

An important discussion of the logical structure of
Carnot's theory which establishes informative parallels
between this and later thermodynamics, exploring in
particular the relationship between "calorique" and
entropy.

1023. Lervig, P., "Sadi Carnot and Nicolas Clément." *Proceedings, 15th International Congress for the History
of Science, Edinburgh, 10-19 August, 1977*, ed. Eric
G. Forbes. Edinburgh: Edinburgh University Press,
1978, pp. 293-304.

A short discussion of Clément's calculation of the
power of the steam engine, and of the role of this
calculation in the flow of ideas in Carnot's 1824
Réflexions.

1024. Mendoza, E., "Contributions to the Study of Sadi
Carnot and His Work." *Arch. Int. Hist. Sci.* 12
(1959): 377-396.

A discussion of a number of questions concerning
Carnot's work, namely: the dating of sources; altera-
tions in MSS; the apparent neglect of Carnot's work;
and on assessing the contemporary significance of his
results in gas physics and engine design.

1025. Mendoza, E., "Sadi Carnot and the Cagnard Engine."
Isis 54 (1963): 262-263.

Suggests that, as conjectured by T.S. Kuhn (item
1021), the Cagnard engine may have influenced Carnot
through the lectures he attended of his friend Nicolas
Clément. Clément used a calculation of the motive
power from a given quantity of heat which seems to
have been based on the mechanism of this engine.

1026. Payen, Jacques, "Un source de la pensée de Sadi Carnot."
Arch. Int. Hist. Sci. 21 (1968): 15-37.

Argues that a course given by Clément-Desormes on
technology including discussion of theories of heat
was an influence on Carnot's thought.

1027. Payen, Jacques, "Deux nouveaux documents sur Nicolas
       Clément." *Rev. Hist. Sci.* 24 (1971): 45-60.

       One of the two documents is an annotated list of
       Clément's works up to 1822 prepared by himself, while
       the other is a table of relative values of work done
       by a steam engine at different temperatures and
       pressures, published in 1826 but of which no copy had
       previously been found. The paper includes commentary
       by Payen.

1028. Redondi, Pietro, "Sadi Carnot et la recherche techno-
       logique en France de 1825 à 1850: présentation d'un
       travail de recherche." *Rev. Hist. Sci.* 29 (1976):
       243-259.

       Argues that there were "echoes" of Carnot's theory in
       the work of later French engineers, particularly
       Franchot and Bresson, and especially in the development
       of air engines.

1029. Redondi, Pietro, *L'accueil des idées de Sadi Carnot
       et la technologie française de 1820 à 1860: de la
       légende à l'histoire.* Paris: Vrin, 1980. 239 pp.

       Based on a fresh and seemingly thorough search of
       the primary sources, this book discusses the intellec-
       tual origins of Carnot's 1824 *Réflexions* and the recep-
       tion of this work prior to its amalgamation with the
       principle of energy conservation around 1850. Reviewed
       by K.R. Hutchison, *Isis* 73 (1982): 145-146.

1030. Rosenfeld, L., "La genèse des principes de la thermo-
       dynamique." *Bull. Soc. Roy. Sci. Liège* 10 (1941):
       199-212.

       An oft-cited paper whose message is now widely known:
       a discussion of the origins of the first and second
       laws of thermodynamics in the work of Carnot, Clapeyron,
       Mayer, Helmholtz, Joule, Clausius and Thomson.

1030a. Russell, C.A., *Time, Chance and Thermodynamics*. Milton
       Keynes: Open University, 1981. 80 pp.

       An Open University text for a course on "Science and
       Belief." Includes sections on the origins of thermo-
       dynamics, Kelvin and the age of the Earth, Maxwell and
       the question of chance or determinism in physics, and
       19th-century reactions to the notion of entropy and
       the concept of the heat-death of the universe.

1031. *Sadi Carnot et l'essor de la thermodynamique.* Paris: Editions du Centre National de la Recherche Scientifique, 1976. 435 pp.

About half of the thirty-five papers contained in this volume are on historical topics, including contributions by Gillispie on Lazare Carnot's influence; Taton on Carnot's education; Birembaut and Herivel on his intellectual milieu; McKeon, Payen and Fox on thermal engineering; Birembaut on Clapeyron; Kastler on Carnot's later work; Mendoza, Schereschewsky, Lervig, Klein and Hoyer on the *Réflexions*; and Kangro and Sadoun-Goupil on later developments in thermodynamics. Essential reading for those interested in the field.

1032. Smith, Crosbie W., "Natural Philosophy and Thermodynamics: William Thomson and 'the Dynamical Theory of Heat.'" *Brit. J. Hist. Sci.* 9 (1976): 293-319.

Describes Thomson's world view as a response to his training in dynamics, his appreciation of the work of Fourier, Joule and Clausius, and his theology, and indicates how changes in this view in order to unite its disparate elements allow us to understand his developing views on thermodynamics.

1033. Smith, Crosbie W., "William Thomson and the Creation of Thermodynamics, 1840-1855." *Arch. Hist. Exact Sci.* 16 (1976-77): 231-288.

Draws on correspondence and unpublished manuscripts to show how Thomson's ideas clarified through a complex process of interaction with his brother James and with Joule, Rankine and Clausius. Argues for greater importance to be attributed to James Thomson in this connection.

1034. Truesdell, C.A., "Absolute Temperature as a Consequence of Carnot's General Axiom." *Arch. Hist. Exact Sci.* 20 (1979): 357-380.

A discussion of thermodynamical axiomatics, centered on the problem of developing an absolute scale of temperatures without using the notion of an ideal gas.

1035. Truesdell, Clifford A., *The Tragicomical History of Thermodynamics, 1822-1854.* New York/Heidelberg/Berlin: Springer-Verlag, 1980. xii + 372 pp.

A rational analysis of the early years of thermodynamics, based on recent developments in the subject.

Though explicitly directed at the scientist rather
than the historian of science, the work contains much
that is also of significance to the latter. Argues
that "thermodynamics need never have been the dismal
swamp of obscurity that from the first it was."
Critical discussion by Philip Lervig, *Centaurus* 26
(1982-83): 85-122.

(iv) Kinetic Theory, Statistical Mechanics, Energeticism

1036. Baucia, Giovanni, "Microfenomeni e macrofenomeni secondo
       Maxwell in relazione alla 'Teoria dinamica del
       calore.'" *Physis* 15 (1973): 333-350.

       Looks at Maxwell's rejection of the applicability of
       dynamical laws to microphenomena in heat theory, his
       demon, and his objection to the interpretations of
       Thomson and Tait, placing this in the context of con-
       temporary philosophical and scientific debates.

1037. Bellone, Enrico, "L'energia molecolare e la velocita
       molecolare come parametri discreti in alcuni scritti
       de L. Boltzmann in relazione all'ipotesi di M. Planck
       ed alla teoria della radiazione di S.D. Poisson."
       Physis 10 (1968): 101-112.

       Suggests that the employment of certain probabilis-
       tic parameters in Boltzmann's gas model used in 1872
       and 1877 may require the "re-reading" of later work.

1038. Bernstein, Henry T., "J. Clerk Maxwell on the History
       of the Kinetic Theory of Gases." *Isis* 54 (1963):
       206-216.

       Transcription with explanatory notes of a letter
       sent by Maxwell to William Thomson (tentatively dated
       to 1871) containing in note form a history of the
       kinetic theory as he saw it, from Lucretius to Hanse-
       mann in 1871.

1038a. Bierhalter, Günter, "Boltzmanns mechanische Grundlegung
       des zweiten Hauptsatzes der Wärmelehre aus dem Jahre
       1866." *Arch. Hist. Exact Sci.* 24 (1981): 195-205.

       Analyzes Boltzmann's 1866 deduction of the second
       law from the kinetic theory, and the subsequent dis-
       cussions by Clausius, Szily and Boltzmann himself of
       the mechanical basis of the demonstration.

1038b. Bierhalter, Günter, "Clausius' mechanische Grundlegung des zweiten Hauptsatzes der Wärmelehre aus dem Jahre 1871." *Arch. Hist. Exact Sci.* 24 (1981): 207-219.

Analyzes Clausius' 1871 memoir in detail in an attempt to distinguish his contributions to the kinetic theory from those of Boltzmann and Szily.

1039. Bierhalter, Günter, "Zu Hermann von Helmholtzens mechanischer Grundlegung der Wärmelehre aus dem Jahre 1884." *Arch. Hist. Exact Sci.* 25 (1981): 71-84.

Summarizes Helmholtz' attempt of 1884 to develop an analogy between the mechanical laws of 'monocyclic' systems and the laws of thermodynamics. Discusses the strengths and weaknesses of Helmholtz' approach. Notes in particular that no mechanical analogue for the law of entropy increase emerges, and that Helmholtz needs to make no assumptions about the particulate structure of matter. Compares this work with earlier attempts by Clausius (1871) and Boltzmann (1866) to provide a mechanical explication of the second law.

1039a. Brock, W.H., ed., *The Atomic Debates: Brodie and the Rejection of the Atomic Theory. Three Studies.* Leicester: Leicester University Press, 1967. x + 186 pp.

Comprises three papers dealing with the skepticism of 19th-century English chemists towards the atomic theory, namely "The Atomic Debates" by W.H. Brock and D.M. Knight (which draws heavily on an earlier paper by the same authors; cf. item 1040); "The Chemical Calculus of Sir Benjamin Brodie" by D.M. Dallas; and "Some Correspondence Connected with Sir Benjamin Brodie's Calculus of Chemical Operations" by W.H. Brock. An appendix discusses the supposed role of Comtean positivism in promoting skepticism about atoms.

1040. Brock, W.H., and D.M. Knight, "The Atomic Debates: 'Memorable and Interesting Evenings in the Life of the Chemical Society.'" *Isis* 56 (1965): 5-25.

The authors argue that while Dalton's atomic theory was widely accepted so far as its explanation of the laws of chemical combination was concerned, up to the 1870s there was grave skepticism as to its ultimate truth, especially with regard to molecular structure.

1041.   Brush, Stephen G., *Kinetic Theory*. 3 vols. Oxford:
        Pergamon Press, 1965-72.

        Each volume provides an introduction followed by
        translations of important papers. Only the first two
        vols. are relevant to our field: Vol. 1, *The Nature
        of Gases and Heat*, containing papers of D. Bernoulli,
        Gregory, Mayer, Joule, Helmholtz, Clausius and Maxwell;
        Vol. 2, *Irreversible Processes*, work by Maxwell, Boltz-
        mann, Thomson, Poincaré and Zermelo.

1042.   Brush, Stephen G., *The Kind of Motion We Call Heat: A
        History of the Kinetic Theory of Gases in the 19th
        Century*. 2 vols. Book 1: *Physics and the Atomists*.
        Book 2: *Statistical Physics and Irreversible Processes*.
        Amsterdam: North-Holland, 1976. xxx + 770 pp.

        This survey of the kinetic theory in the 19th century
        is fundamentally a collection of 15 of the author's
        previously published papers. To these papers have
        been added a 100-page introductory chapter, a thorough
        bibliography of primary sources, minor modifications
        called for by Brush's ongoing research, and various
        additional passages designed to increase the coherence
        of the work. The book is concerned principally with
        technical matters internal to the kinetic theory,
        though there is significant discussion of the intellec-
        tual context within which the theory developed. Chapters
        discuss Herapath, Waterston, Clausius, Maxwell, Boltz-
        mann, Van der Waals, and Mach (in Book I); then the
        wave theory of heat, foundations of statistical mech-
        anics 1845-1915, interatomic forces and the equation
        of state, viscosity theory, heat conduction and the
        Stefan-Boltzmann law, irreversibility, and Brownian
        motion (in Book II). Undoubtedly a major secondary
        source. See the reviews of E. Mendoza, *Ann. Sci.* 35
        (1978): 332-333, and R.H. Stuewer, *Isis* 69 (1978):
        137-138. Cf. item 954.

1043.   Brush, Stephen G., "Irreversibility and Indeterminism:
        Fourier to Heisenberg." *J. Hist. Ideas* 37 (1976):
        603-630.

        Looking primarily at the notion of irreversibility
        in the second law of thermodynamics, Brush sketches
        out the gradual acceptance, especially by Maxwell and
        Boltzmann, of the physical reality of randomness, over
        and above the acceptance of it as a calculational
        device.

1044. Clark, Peter, "Atomism versus Thermodynamics." *Method and Appraisal in the Physical Sciences: The Critical Background to Modern Science, 1800-1905*, ed. Colin Howson. Cambridge: Cambridge University Press, 1976, pp. 41-105.

Adopting a Lakatosian model, divides 19th-century theorizing about heat into (i) an atomic-kinetic research programme based on a specific theory of matter (Maxwell, Boltzmann) and (ii) a thermodynamic research programme drawn from the general mechanical theory of heat and developing into energetics (Clausius, Gibbs, Planck). Traces the degeneration of the kinetic research programme (1880-1905) and its later regeneration. A general criticism of the possible limitations of this type of approach is in G. Gutting's review, *Isis* 69 (1978): 309-310.

1045. Daub, Edward E., "Probability and Thermodynamics: The Reduction of the Second Law." *Isis* 68 (1969): 318-330.

By working through Boltzmann's successive systems, Daub shows how probability, though alien to mechanics, was incorporated into the reduction of the second law, and indicates its importance in the later debate between energetics and atomism in thermodynamics.

1046. Daub, Edward E., "Waterston, Rankine, and Clausius on the Kinetic Theory of Gases." *Isis* 61 (1970): 105-106.

Shows, *contra* Brush (*Ann. Sci.* 13 (1957): 273-282) that Waterston's work was in fact presented in a paper by Rankine in 1853, and was thus not completely unknown.

1047. Daub, Edward E., "Maxwell's Demon." *Studs. Hist. Phil. Sci.* 1 (1970): 213-227.

Discusses the problem of interpreting the second law of thermodynamics which led to the conviction that this law was not reducible to purely mechanical phenomena.

1048. Daub, Edward E., "Waterston's Influence on Krönig's Kinetic Theory of Gases." *Isis* 62 (1971): 512-515.

Establishes the possibility of a link between Waterston and Krönig by comparing their papers and by pointing to a means (the journal Krönig edited) by which the

ideas could have been transmitted, thus increasing the
significance of Waterston's work and indicating how
it took root in Germany.

1049.   Dorling, Jon, "Maxwell's Attempt to Arrive at Non-
        Speculative Foundations for the Kinetic Theory."
        Studs. Hist. Phil. Sci. 1 (1970): 229-248.

        An evaluation of methodological issues involved in
        Maxwell's attempts to base the kinetic theory on in-
        ductively reasonable evidence rather than on naive
        hypothetico-deductive reasoning.

1050.   Dugas, René, *La théorie physique au sens de Boltzmann
        et ses prolongements modernes.* Neuchâtel: Editions
        du Griffon, 1959.   308 pp.

        A detailed exposition of Boltzmann's attitude towards
        physical theory, illustrated by an analysis of his
        H-theorem and the kinetic interpretation of entropy.
        This is followed by a somewhat briefer discussion of
        subsequent work in statistical mechanics by Gibbs,
        Poincaré, Planck and Einstein.

1051.   Elkana, Yehuda, "Boltzmann's Scientific Research Pro-
        gram and Its Alternatives." *The Interaction Between
        Science and Philosophy*, ed. Yehuda Elkana. Atlantic
        Highlands, N.J.: Humanities Press, 1974, pp. 243-
        279.

        Written in Lakatosian language, this article follows
        Boltzmann from his advocacy of an atomist-mechanist
        research program in the 1860s, to his adoption in his
        last years of the view that scientific entities were
        mere "mental pictures."

1052.   Garber, Elizabeth, "Maxwell, Clausius and Gibbs: As-
        pects of the Development of Kinetic Theory and
        Thermodynamics." Ph.D. dissertation, Case Institute
        of Technology, 1966.

        Examines the correspondence between Maxwell and
        Clausius concerning the former's innovations in kinetic
        theory, arguing that although initially fruitful it
        eventually dwindled as neither understood the other's
        method.   This can be seen in Maxwell's only appreciating
        the potential of thermodynamics twenty years after
        Clausius' important work, when he read Gibbs' papers
        of 1873 and 1876 which were in method similar to his
        own "geometrical thought."

1053. Garber, Elizabeth, "Clausius and Maxwell's Kinetic Theory of Gases." *Hist. Studs. Phys. Sci.* 2 (1970): 299-319.

Shows that Clausius contributed the valuable ideas of mean free path and statistical analyses to Maxwell's early gas theory. Shows, too, how his criticisms shaped Maxwell's work on distribution functions and transport theory, even though the methods of the two men increasingly diverged.

1054. Garber, Elizabeth, "Aspects of the Introduction of Probability into Physics." *Centaurus* 17 (1972): 11-39.

Argues that prior to Maxwell there was little use of probability in physics except in data analysis, but that Maxwell was aware even in his early work of the need for a statistical treatment of gas theory, drawing "casually" on the model of error distribution theory in order to develop this.

1055. Garber, Elizabeth, "Molecular Science in Late-Nineteenth-Century Britain." *Hist. Studs. Phys. Sci.* 9 (1978): 265-298.

Describes how work on kinetic theory led (*c.* 1860-1900) to the inclusion of molecular theorizing into British physics, how all the many models evolved subsequent to Maxwell's work had the same fundamental problems, but how nonetheless the same foundation was always tacitly assumed.

1056. Gillispie, C.C., "Intellectual Factors in the Background of Analysis by Probabilities." *Scientific Change*, ed. A.C. Crombie. London: Heinemann, 1963, pp. 431-453.

Seeks out the intellectual antecedents of Maxwell's inauguration in 1859 of the science of statistical mechanics. Finds these in applications of probability in various 19th-century writings on "social physics," and especially in the work of Quetelet and, after him, John Herschel. Mary B. Hesse in her printed comments (pp. 471-476) disputes this and asserts anew the seminal importance for Maxwell of Clausius' discussion of mean free paths of molecules in gases. A reply by Gillispie is also included (pp. 499-502).

1057.   Heimann, P.M., "Molecular Forces, Statistical Repre-
        sentation and Maxwell's Demon." *Studs. Hist. Phil.
        Sci.* 1 (1970): 189-211.

        Discusses a number of issues: Maxwell's introduction
        to probabilistic reasoning and his attitude to statis-
        tical theories in physics; the influence of Clausius;
        the nature of matter in Maxwell's work; his distinctions
        between macro and micro-phenomena.

1058.   Helm, Georg, *Die Energetik nach ihrer geschichtlichen
        Entwickelung.* Leipzig, 1898. Reprinted, New York:
        Arno Press, 1981. xii + 370 pp.

        A classic work, written from the energeticist point
        of view developed by Ostwald and his school.

1059.   Hiebert, Erwin, "The Energetics Controversy and the
        New Thermodynamics." *Perspectives in the History of
        Science and Technology*, ed. Duane H.D. Roller.
        Norman: University of Oklahoma Press, 1972, pp.
        67-97.

        Essentially a summary of the positions taken in the
        short, sharp conflict between non-materialist 'ener-
        getics' and molecular-kinetic physics about 1895,
        showing that there was minimal debate prior to that
        date.

1060.   Hiebert, Erwin N., "Boltzmann's Conception of Theory
        Construction: The Promotion of Pluralism, Provisional-
        ism, and Pragmatic Realism." *Probabilistic Thinking,
        Thermodynamics and the Interaction of the History
        and Philosophy of Science: Proceedings of the 1978
        Pisa Conference on the History and Philosophy of
        Science, Vol. II*, ed. Jaakko Hintikka, David Gruender
        and Evandro Agazzi. Dordrecht/Boston: D. Reidel,
        1981, pp. 175-198.

        Surveys Boltzmann's writings, from the latter period
        of his life, on the epistemological status of scientific
        theories. Suggests that Boltzmann's philosophy of
        science ought to be seen as an attempt to justify his
        own way of doing science.

1061.   Holt, Niles R., "A Note on Wilhelm Ostwald's Energism."
        Isis 61 (1970): 386-389.

        Briefly recounts Ostwald's attempts (1887-1932) to
        discredit atomism, and his gradually expanding concept

of Energism--the idea that atomic particles were mathematical fictions used to explain the operations of
energy--that he proposed as an alternative. Includes
a good primary bibliography.

1061a. Hoyer, Ulrich, "Von Boltzmann zu Planck." *Arch. Hist.
Exact Sci.* 23 (1980): 47-86.

Discusses the connections between Planck's quantum
theory of radiation and Boltzmann's statistical conceptions, analyzing the latter in some detail in the
process. Concludes that while, historically speaking,
Planck's notion of finite energy elements was a revolutionary one, nevertheless from a systematic standpoint
his approach was conservative and classical.

1062. Kartsev, V., "The Mach-Boltzmann Controversy and Max-
well's Views on Physical Reality." *Probabilistic
Thinking, Thermodynamics and the Interaction of the
History and Philosophy of Science: Proceedings of
the 1978 Pisa Conference on the History and Philosophy
of Science, Vol. II*, ed. Jaakko Hintikka, David
Gruender and Evandro Agazzi. Dordrecht/Boston:
D. Reidel, 1981, pp. 199-205.

Argues that "Boltzmann consciously accepted Maxwell's
methodology and fought against attempts to regard Maxwell as a phenomenologist."

1063. Klein, Martin J., "Maxwell, His Demon, and the Second
Law of Thermodynamics." *Amer. Scientist* 58 (1970):
84-97.

Discusses Maxwell's appreciation of the statistical
nature of the second law, Boltzmann's and Clausius'
contemporary efforts to find an adequate mechanical
law, and the ultimate success of the statistical approach with Boltzmann's later work.

1064. Klein, Martin J., "Mechanical Explanation at the End
of the Nineteenth Century." *Centaurus* 17 (1972):
58-82.

Discusses the search for a mechanical explanation of
the irreversibility implied by the second law of
thermodynamics. Centres on Helmholtz' 1884 theory
of monocyclic systems, and the responses of Hertz and
Boltzmann to this theory.

1065. Klein, M.J., "The Development of Boltzmann's Statis-
       tical Ideas." *The Boltzmann Equation: Theory and
       Applications*, ed. E.G.D. Cohen and W. Thirring.
       Wien/New York: Springer-Verlag, 1973. (Acta Physica
       Austriaca, Suppl. 10), pp. 53-106.

       Traces the development of Boltzmann's ideas on the
       second law of thermodynamics from his early efforts to
       provide a purely mechanical explanation to his eventual
       acceptance of the need to introduce statistical con-
       cepts. The impact on his thinking of criticisms
       levelled at his work at various times is analyzed in
       detail, as is the degree to which his work was under-
       stood and accepted by his contemporaries.

1066. Klein, Martin J., "Boltzmann, Monocycles and Mechanical
       Explanation." *Boston Studies in the Philosophy of
       Science, Vol. 11: Philosophical Foundations of
       Science*, ed. R.J. Seeger and R.S. Cohen. Dordrecht/
       Boston: D. Reidel, 1974, pp. 155-175.

       Argues that the dominant goal of many 19th-century
       physicists was the mechanical explanation of phenomena,
       and looks at the influence of Helmholtz' analogy of
       monocycles for the second law of thermodynamics on the
       work of Hertz and especially Boltzmann.

1067. Klein, M.J., "The Historical Origins of the Van der
       Waals Equation." *Physica* 73 (1974): 28-47.

       Analyzes Van der Waals' ideas on molecular theory
       as set out in his 1873 thesis, and also the status of
       the theory more generally at that time. Traces the
       influence on Van der Waals of Clausius in particular.

1068. Knight, David M., "Steps towards a Dynamical Chemistry."
       Ambix 14 (1967): 179-197.

       Shows that the chemically inclined scientists of the
       nineteenth century increasingly favoured a dynamical
       theory of matter against a "mechanical" particulate
       theory.

1069. Knight, David M., *Atoms and Elements: A Study of
       Theories of Matter in England in the Nineteenth
       Century*. London: Hutchinson, 1967. 167 pp.

       Concentrates more on the developments and problems
       of chemical theories of matter, but also contains
       some useful summaries of various physical atomic
       theories.

1070. Krajewski, W., "The Idea of Statistical Law in Nine-
teenth Century Science." *Boston Studies in the
Philosophy of Science, Vol. 14: Methodological and
Historical Essays in the Natural and Social Sciences*,
ed. R.S. Cohen and M.W. Wartofsky. Dordrecht/Boston:
D. Reidel, 1974, pp. 397-405.

Briefly traces the acceptance of statistical laws by
the scientific community, firstly as empirical derived
relationships in the social and biological sciences,
and then in theoretical derivations from the "summa-
tions" of idealized individual cases in the physics
of Maxwell and Boltzmann.

1071. Lindsay, Robert Bruce, ed., *Early Concepts of Energy
in Atomic Physics*. (Benchmark Papers on Energy, vol.
7.) Stroudsburg, Penn.: Dowden, Hutchinson and Ross,
1979. xiv + 402 pp.

Reprints of papers (mostly 20th-century) divided into
sections chronologically and thematically, with a
brief historical introduction to each section by the
editor. Includes extracts on kinetic theory by Davy,
Herapath, Waterston, Daniel Bernoulli, Boscovich and
Brown. A section on the "Energetic Description of
Atomic Motion" includes papers by Joule, Krönig,
Clausius and Maxwell.

1072. Mendoza, Eric, "The Surprising History of the Kinetic
Theory of Gases." *Mem. Proc. Manchester Lit. Phil.
Soc.* 105 (1962-63): 15-28.

Outlines the early kinetic theories of Herapath
and Joule, and the difficulties facing those theories.
An important part of the discussion has been superseded
by a later paper by the same author (item 1073).

1073. Mendoza, E., "A Critical Examination of Herapath's
Dynamical Theory of Gases." *Brit. J. Hist. Sci.* 8
(1975): 155-165.

Argues that Herapath merely modified the accepted
static lattice theory of gases into a vibrating lattice
model, and did not create a fully kinetic theory of
gases.

1074. Nye, Mary Jo, *Molecular Reality: A Perspective on the
Scientific Work of Jean Perrin*. London: Macdonald/
New York: American Elsevier, 1972. xii + 201 pp.

Begins with an account of late 19th-century debates, especially in France, between those who maintained the physical reality of atoms and molecules, and "energeticists" and others who regarded them as mere mental constructs or worse. Perrin's work in the early years of the 20th century on Brownian motion and related topics is then discussed in detail, this being seen as part of a conscious (and successful) attempt to provide direct empirical support for the molecular theory.

1075.   Nye, Mary Jo, "The Nineteenth-Century Atomic Debates and the Dilemma of an 'Indifferent Hypothesis.'" *Studs. Hist. Phil. Sci.* 7 (1976): 245-268.

Shows how empirical criticisms caused atomism to become (1860-1895) an "indifferent hypothesis," neither validated nor invalidated, and examines the anti-reductionist, anti-mechanist criticisms of the advocates of the alternative thermodynamic approach who appealed for a "purely pragmatic methodology of science."

1076.   Paul, E. Robert, "Alexander W. Williamson on the Atomic Theory: A Study of Nineteenth-Century British Atomism." *Ann. Sci.* 35 (1978): 17-31.

Shows how Williamson, influenced by Leibig's anti-*Naturphilosophie* outlook and his atomism, and by Comte's positivism, evolved an influential atomic theory to explain chemical affinity.

1077.   Porter, Theodore M., "A Statistical Survey of Gases: Maxwell's Social Physics." *Hist. Studs. Phys. Sci.* 12 (1981): 77-116.

Argues that Maxwell's discovery of the molecular velocity distribution for gases was an inference from contemporary statistical thought, and involved no connotation of uncertainty. His later discovery of statistical uncertainty in physics "represented an important transformation of European statistical thinking"; it arose, it is suggested, in large measure "from Maxwell's feelings about the proper relation between science and religious truth."

1078.   Schneider, Ivo, "Clausius' erste Anwendung der Wahrscheinlichkeitsrechnung im Rahmen der atmosphärischen Lichtstreuung." *Arch. Hist. Exact Sci.* 14 (1974-75): 143-158.

Shows that Clausius and not Maxwell was the first to
introduce probabilistic methods into physics, using
the theory of probability to eliminate a hypothesis in
early work on metereological optics, as well as to
justify his later work (1849) on mean values to
describe large numbers of particles. Suggests that
he probably derived his ideas from Lambert and possibly
Young.

1079.   Schneider, Ivo, "Rudolph Clausius' Beitrag zur Ein-
        fuhrung wahrscheinlichkeitstheoretischer Methoden in
        die Physik der Gase nach 1856." *Arch. Hist. Exact
        Sci.* 14 (1974-75): 237-260.

Describes the development of probability theory in
the kinetic theory of gases. Includes Laplace's original
initiative on the possibilities of probability theory
leading to Krönig's work (1856) reducing a multitude
of atoms to six-atom groups, Clausius' introduction
(1858) of mean free path and Maxwell's formulation
(1859) of the molecular velocity distribution.

*   Truesdell, C.A., "Early Kinetic Theories of Gases."

        Cited herein as item 337.

1080.   Wilson, David B., "Kinetic Atom." *Amer. J. Phys.* 49
        (1981): 217-222.

Surveys mid-to-late 19th-century atomic theory including the work of Stokes, Kelvin, Maxwell amd Boltzmann, arguing that attitudes to theories were largely
nationally determined but were also cautious.

                (h) Electricity and Magnetism

(i) Early Electrodynamics

1081.   Blondel, Christine, "Sur les premières recherches de
        formule électrodynamique par Ampère (octobre 1820)."
        Rev. Hist. Sci. 31 (1978): 53-65.

Publishes two MSS showing that as early as October
1820, within weeks of hearing of Oerstad's discovery,
Ampère was seeking a quantitative expression for the
force between two current elements.

1082.   Brown, Theodore, "The Electric Current in Early Nine-
        teenth-Century French Physics." *Hist. Studs. Phys.
        Sci.* 1 (1969): 61-103.

        Describes the theory of the Voltaic pile and electric
        current from Volta in 1800 to Biot's alternative theory
        using Coulomb's electrostatics; the dynamic "reunion"
        theory of the "new" scientists Ampère, Becquerel and
        De la Rive which was initiated by Oersted's discovery
        of electromagnetism; and a final assimilation of the
        last two by Pouillet (1828).

*   Caneva, Kenneth L., "From Galvanism to Electrodynamics:
        The Transformation of German Physics and Its Social
        Context."

        Cited herein as item 758.

1083.   Caneva, Kenneth L., "Ampère, the Etherians, and the
        Oersted Connexion." *Brit. J. Hist. Sci.* 13 (1980):
        121-138.

        Seeks to establish the reasons for Ampère's immediate
        and positive response to Oersted's discovery of electro-
        magnetism. Finds these in his firm attachment to a
        world-view in which the aether played a unifying con-
        ceptual role--this putting him in the anti-Laplacian
        camp within the French scientific community of the
        time--and in his acquaintance with and respect for
        Oersted's earlier work.

1083a. Caneva, Kenneth L., "What Should We Do with the Monster?:
        Electromagnetism and the Psychosociology of Knowledge."
        *Sciences and Cultures: Anthropological and Historical
        Studies of the Sciences*, ed. Everett Mendelsohn and
        Yehuda Elkana. Dordrecht: D. Reidel, 1981. pp. 101-
        131.

        An interesting attempt to apply Mary Douglas' social
        anthropological theory of group and grid to the early
        history of electromagnetism, arguing that the responses
        of different physicists to Oersted's discovery lead to
        a classification of these physicists according to the
        group-grid theory that is generally consistent with the
        other things we know about them.

*   Cawood, John A., "The Scientific Work of D.F.J. Arago,
        1786-1853."

        Cited herein as item 714.

1084.   Coutts, A., "William Cruickshank of Woolwich." *Ann.
        Sci.* 15 (1959): 121-133.

Describes the life and work of Cruickshank (?-c. 1811), confused by a number of writers with the better known William Cumberland Cruickshank. Includes a description of Cruickshank's very early work on electrolysis and his innovations in Voltaic pile design (1801).

*   Fuchtbauer, Heinrich von, *Georg Simon Ohm: Ein Forscher wachst aus seiner Vater Art.*

Cited herein as item 51.

1085. Gardiner, K.R., and D.L. Gardiner, "André-Marie Ampère and his English Acquaintances." *Brit. J. Hist. Sci.* 2 (1965-66): 235-245.

Shows that there was an extensive and amiable correspondence between Ampère and English scientists such as Davy, John Herschel, Babbage and especially Faraday.

1086. Hamamdjian, Pierre-Gérard, "Genèse des idées d'Ampère en électromagnetisme." *Proceedings, 12th International Congress of the History of Science, Paris, 1968* (Paris, 1971), vol. 5, pp. 29-34.

Argues that even prior to Oersted's discovery Ampère had rejected magnetic fluids as separate substances and was susceptible to an electrical theory of magnetism.

1087. Hamamdjian, Pierre-Gérard, "Contribution d'Ampère au 'théoreme d'Ampère.'" *Rev. Hist. Sci.* 31 (1978): 249-268.

Using an unpublished MS argues that "Ampère's theorem" could never have been derived by Ampère due to the restrictions of his physical theory of magnetism. Compares Ampère's work in the MS with Maxwell's later treatment of the same problem.

*   Heidelberger, Michael, "Some Patterns of Change in the Baconian Sciences of Early 19th Century Germany."

Cited herein as item 766. Chiefly concerned with Ohm's work on electricity.

1088. Hofmann, James R., "The Great Turning Point in André-Marie Ampère's Research in Electrodynamics: A Truly 'Crucial' Experiment." Ph.D. dissertation, University of Pittsburgh, 1982.

A detailed study, based on extensive archival investigations, of the roots of Ampère's technique of using "null" experiments as the basis for his derivation of the electrodynamic force law. Stresses the role of

one experiment in particular, performed in 1822, in
the development of Ampère's thinking.

1089. Humphreys, A.W., "The Development of the Conception and
       Measurement of Electric Current." *Ann. Sci.* 2
       (1937): 164-178.

       Points out some physical problems in early measure-
       ments of currents (lack of a steady source of EMF, the
       internal resistance of cells, lack of knowledge about
       galvanometers), indicates the conceptual vagueness
       surrounding the very notion of current, and briefly
       describes the attempts to resolve these problems, *c.*
       1820-1840.

1090. Kastler, Alfred, "Ampère et les lois de l'électrody-
       namique." *Rev. Hist. Sci.* 30 (1977): 143-157.

       An exposition of Ampère's route to his law of force
       between current elements, summarizing Ampère's own
       account in his *Mémoire sur la théorie mathématique
       des phénomènes électrodynamiques, uniquement déduite
       de l'expérience.*

1091. King, W. James, "The Quantification of the Concept of
       Electric Charge and Electric Current, Part I."
       Natural Philosopher 2 (1963): 107-127.

       A brief survey of developments up to and including
       the work of Weber.

1092. Lezhneva, O.A., "Electricity and Magnetism in Russia
       in the Middle of the XIXth Century." *Actes du VIIIe
       Congrès International d'Histoire des Sciences,
       Florence-Milan, 1956* (Florence, 1958), vol. 1, pp.
       289-292.

       Concentrates on the work of Lenz and Jacobi.

*   Lezhneva, O.A., "Die Entwicklung der Physik in Russ-
       land in der ersten Hälfte des 19. Jahrhunderts."

       Cited herein as item 812. Chiefly concerned with
       the electrical studies of Petrov and Lenz.

1093. McKnight, John L., "Laboratory Notebooks of G.S. Ohm:
       A Case Study in Experimental Method." *Amer. J. Phys.*
       35 (1967): 110-114.

       A brief description of the intermeshing of theory
       and experiment in Ohm's work.

1094. Merleau-Ponty, Jacques, *Leçons sur la genèse des
       théories physiques: Galilée, Ampère, Einstein.*
       Paris: Vrin, 1974. 172 pp.

Includes (pp. 69-112) an interesting discussion of
Ampère's *Théorie mathématique des phénomènes électro-
dynamiques* (1827), viewing this as illustrative of
one of the characteristic stages in the formation of
modern physical theory in which the Galilean objective
of mathematizing experimental knowledge is accepted,
but in which a new problem arises where, with this
objective in mind, a choice has to be made between
different modes of conceptualizing the situation.

1095. Olson, Richard G., "Sir John Leslie and the Laws of
        Electrical Conduction in Solids." *Amer. J. Phys.* 37
        (1969): 190-195.

Contends that Leslie provided both a theoretical discussion and a limited experimental confirmation of
Ohm's Law in a paper written in 1791 and published in
1824, three years prior to Ohm's presentation.

1096. Ross, Sydney, "The Search for Electromagnetic Induc-
        tion, 1820-1831." *Notes Rec. Roy. Soc. Lond.* 20
        (1965): 184-219.

Describes various efforts to produce electricity
from magnetism, and the discovery of various related
phenomena such as electromagnetic rotation, between
1820 and 1831. A large part of the paper is devoted
to the work of Ampère and Faraday and their co-operative
but largely unproductive correspondence. Also attempts
to explain why Ampère did not follow up his discovery
of a "clue" to the problem, and describes the reaction
to Faraday's eventual success.

1097. Russell, Colin A., "The Electrochemical Theory of Sir
        Humphry Davy. I: The Voltaic Pile and Electrolysis."
        Ann. Sci. 15 (1959): 1-13.

Describes Davy's theory of the Voltaic pile based on
both the "chemical theories" in which chemical action
was seen as necessary to electrical generation, and,
though less so, on "contact" theories.

1098. Schagrin, Morton L., "Resistance to Ohm's Law." *Amer.
        J. Phys.* 31 (1963): 536-547.

Provides a good account of Ohm's early experiments
leading to the discovery of his law (1825-1827).
Describes the concepts of "current" and "tension"
generally seen at the time to be involved in the opera-
tion of the electric cell, and argues that Ohm's work
was largely ignored because of an "incomprehensible"
conceptual shift which ignored the accepted distinction
between these two concepts.

1099.   Snelders, H.A.M., "The Reception in the Netherlands of
        the Discoveries of Electromagnetism and Electrodynamics
        (1820-1822)." *Ann. Sci.* 32 (1975): 39-54.

        Describes in its theoretical context Dutch experimen-
        tal work, especially that of van Beek and his circle,
        in repeating, extending and eventually confirming the
        experiments of Oersted, Arago and Ampère.

1100.   Spencer, J. Brookes, "Boscovich's Theory and its Rela-
        tion to Faraday's Researches: An Analytic Approach."
        Arch. Hist. Exact Sci. 4 (1967-68): 184-202.

        Argues *contra* Williams (item 105) that there is no
        significant correspondence between the systems of
        Faraday and Boscovich.

1101.   Stauffer, R.C., "Persistent Errors Regarding Oersted's
        Discovery of Electromagnetism." *Isis* 44 (1953):
        307-310.

        Shows that Oersted first postulated a link between
        electricity and magnetism in 1812 and devised an ex-
        periment to test it in 1820. Immediately, an error in
        dating was published by J.N.P. Hachette, while another
        mistake, the allegation that the discovery was acci-
        dental, was perpetrated by L.W. Gilbert, the editor of
        Annalen der Physik.

1102.   Stauffer, R.C., "Speculation and Experiment in the
        Background of Oersted's Discovery of Electromagnetism."
        Isis 48 (1957): 33-50.

        Argues that Oersted's discovery of electromagnetism
        was the product of long speculation and experimenta-
        tion, partly stimulated by his knowledge of and agree-
        ment with the teachings of *Naturphilosophie* and of
        his friend Ritter on the unity of the forces of nature.

1103.   Stine, Wilbur Morris, *The Contributions of H.F.E. Lenz
        to Electromagnetism*. Philadelphia: The Acorn Press,
        1923. 157 pp.

        Includes a very sketchy biographical chapter, together
        with a bibliography of Lenz' scientific publications and
        an account, unfortunately historically exceedingly naive
        and with little mathematical detail, of his major papers
        on electromagnetism.

*   Sutton, Geoffrey, "The Politics of Science in Early
          Napoleonic France: The Case of the Voltaic Pile."
          Cited herein as item 749.

1104.   Tricker, R.A.R., "Ampère as a Contemporary Physicist."
          Contemporary Physics 3 (1960-61): 453-468.

          Describes the experiments involved in Ampère's
          derivation (1820-26) of his law of action between
          electric current elements, examines criticisms of the
          law by Grassmann, Heaviside and Whittaker, and compares
          it with the less consistent but simpler contemporary
          Biot-Savart law.

1105.   Tricker, R.A.R., *Early Electrodynamics: The First Law
          of Circulation.* Oxford: Pergamon, 1965. x + 217 pp.

          Deals with the development of Ampère's theory of
          the electrodynamics of steady currents. Includes ex-
          tensive extracts from Ampère's writings and those of
          Biot and Savart, shorter ones from papers by Oersted
          (1820) and Grassmann (1845), two short chapters setting
          the historical stage, and an 89-page commentary. The
          latter is not purely historical, but is also concerned
          with elucidating the logical status of the theory.

1106.   Tunbridge, Paul A., "Faraday's Genevese Friends."
          Notes Rec. Roy. Soc. Lond. 27 (1972-73): 263-298.

          Drawing mainly on their correspondence, sketches
          the highly amicable relationship of Faraday with first
          Gaspard and then his son Auguste De la Rive. Topics
          covered in the letters include Faraday's brief priority
          dispute with Ampère over electromagnetic induction,
          paramagnetism, Faraday's thoughts on light and magnetism
          (1846), Auguste's matter theory (1854), and a discussion
          of electrical discharges in gases (c. 1858).

1107.   Williams, L. Pearce, "Ampère's Electrodynamical Molecu-
          lar Model." *Contemporary Physics* 4 (1962-63): 113-
          123.

          Describes the evolution of Ampère's highly compli-
          cated molecular model out of his theory of electric
          current, in which the molecule was seen as a miniature
          voltaic pile and by which he attempted to account for
          electromagnetic phenomena.

1108. Williams, L. Pearce, "The Simultaneous Discovery of
       Electro-Magnetic Induction by Michael Faraday and
       Joseph Henry." *Bull. Soc. Amis A.-M. Ampère* no. 22
       (Jan. 1965): 12-21.

       Not seen.

*   Williams, L. Pearce, *Michael Faraday: A Biography*.

       Cited herein as item 105.

1108a. Winter, H.J.J., "The Significance of the Bakerian Lecture
       of 1843." *Philosophical Magazine* ser. 7, 34 (1943):
       700-711.

       A discussion of Wheatstone's paper on electrical
       circuits that provided important support for Ohm's
       ideas, especially by offering improved methods of
       measuring resistance.

1109. Winter, H.J.J., "The Work of G.T. Fechner on the Gal-
       vanic Circuit." *Ann. Sci.* 6 (1948-50): 197-205.

       Describes Fechner's experiments used to provide the
       first verification of Ohm's law (1831) and his indepen-
       dent discovery of the formulae for simple and parallel
       circuits, and describes the difficulties involved in
       getting reliable results.

1110. Yamazaki, Eizo, "Sur l'électrodynamique d'Ampère."
       Mem. Inst. Sci. Tech. Meiji Univ. 16(12) (1977):
       1-31.

       Analyzes in some detail the evolution of Ampère's
       ideas on electrodynamics, arguing that, though he was
       an excellent experimenter, his outlook was above all
       that of a theoretician, his primary achievement that
       of mathematizing what had been until then a qualitative
       and experimental branch of physics.

(ii) Electromagnetism, Field Theory

*   Agassi, Joseph, *Faraday as a Natural Philosopher*.

       Cited herein as item 28.

1111. Agassi, Joseph, "Field Theory in De la Rive's *Treatise
       on Electricity*." *Organon* 11 (1975): 285-301.

       Chiefly concerned with De la Rive's presentation of
       Faraday's ideas.

1112. d'Agostino, Salvo, "I vortici dell'etere nella teoria del campo elettromagnetico di Maxwell: la funzione del modello nella costruzione della teoria." *Physis* 10 (1968): 188-202.

Discusses the different roles played by mechanical models of the aether in Maxwell's various papers on electrodynamics.

1113. d'Agostino, Salvo, "Il pensiero scientifico di Maxwell e lo sviluppo della teoria del campo elettromagnetico nella memoria 'On Faraday's Lines of Force.'" *Scientia* 103 (1968): 291-301. (French translation in supplement, pp. 155-164.)

Examines Maxwell's ideas on the nature of physical theories and on methodology, as set out in his first memoir on electromagnetism and elsewhere. Focuses particularly on his advocacy of theoretical pluralism, and on his search for mechanical models for physical theories.

1114. d'Agostino, Salvo, "Hertz's Researches on Electromagnetic Waves." *Hist. Studs. Phys. Sci.* 6 (1975): 261-323.

An extended discussion of the development during the 1880s of Hertz' theoretical conceptions in electrodynamics leading to his famous experiments on electric waves. The importance for Hertz of Helmholtz' theory of polarization is emphasized, but so, too, is his ultimate acceptance of a purer Maxwellian outlook.

1115. d'Agostino, Salvo, "La scopertà di una velocità quasi uguale alla velocità della luce nell'elettrodinamica di Wilhelm Weber (1804-1891)." *Physis* 18 (1976): 297-318.

Describes how a constant with the dimensions of a velocity appeared in Weber's fundamental law of electrodynamic action, and discusses the implications of the discovery by Weber and Kohlrausch that its value approximated that of the velocity of light. Argues that Weber did not attach any special significance to this result, and in this connection contrasts Weber's research program, seen as aiming to reduce electrodynamics to mechanics, with Maxwell's program based on contiguous action and the notion of a field.

1116. d'Agostino, Salvo, "Weber and Maxwell on the Discovery
       of the Velocity of Light in Nineteenth Century
       Electrodynamics." *On Scientific Discovery: The
       Erice Lectures 1977*, ed. M.D. Grmek, R.S. Cohen and
       G. Cimino. Dordrecht/Boston: D. Reidel, 1981, pp.
       281-293.

       Contrasts the role of the velocity factor in the
       theories of Weber and Maxwell, while stressing the
       importance of metrological considerations to both men.
       Treats this as a case study in relation to philosophical
       questions about overlapping empirical content between
       different scientific theories.

1117. d'Agostino, Salvo, "Esperimento e teoria nell'opera di
       Maxwell: Le misure per le unità assolute elettromag-
       netiche e la velocità della luce." *Scientia* 113
       (1978): 453-468; English translation pp. 469-480.

       Suggests that there is a correlation between Max-
       well's experimental activity as a member of the British
       Association's committee for the determination of elec-
       trical standards and the successive stages in the formu-
       lation of his electromagnetic theory.

1118. Berkson, William, *Fields of Force: The Development of
       a World View from Faraday to Einstein*. London:
       Routledge and Kegan Paul, 1974. xiv + 370 pp.

       A lively survey of the development of the concept
       of a field from Faraday to Einstein. Is particularly
       concerned to delineate the "problem situation" within
       which each person discussed (most notably Faraday,
       Maxwell, Hertz, Lorentz and Einstein) was working.

1119. Bork, Alfred M., "Maxwell, Displacement Current, and
       Symmetry." *Amer. J. Phys.* 31 (1963): 854-859.

       Argues against the statement often made in physics
       texts that Maxwell introduced the displacement current
       to make his equations symmetrical. Shows that sym-
       metry was brought in as a consideration by Heaviside.

1120. Bork, Alfred M., "Maxwell and the Electromagnetic Wave
       Equations." *Amer. J. Phys.* 35 (1967): 844-849.

       Discusses and illustrates by flow charts Maxwell's
       three derivations of the wave equation from the basic
       electromagnetic equations, in the papers of 1865 and
       1868 and in the *Treatise*.

1121. Bork, Alfred M., "Maxwell and the Vector Potential."
    Isis 58 (1967): 201-222.

    Presents detailed evidence to show that, in contrast
    to a commonly expressed modern view, for Maxwell the
    vector potential represented not a mere mathematical
    construction but a real physical quantity which had
    its conceptual origin in Faraday's notion of the
    "electrotonic state."

1122. Bromberg, Joan, "Maxwell's Displacement Current and His
    Theory of Light." *Arch. Hist. Exact Sci.* 4 (1967):
    218-234.

    Rejects, on the basis of a detailed analysis of the
    relevant parts of Maxwell's published papers, the
    accounts usually given of his introduction of the dis-
    placement current, arguing that he introduced it not
    on aesthetic grounds or in order to obtain a consistent
    set of equations but for pragmatic reasons, as a plausible
    method of advancing his calculations. Points to errors
    and serious confusions in Maxwell's initial presentation
    of the idea, and shows how these were gradually elimi-
    nated in his later publications on electromagnetism.

1123. Bromberg, Joan, "Maxwell's Electrostatics." *Amer. J.
    Phys.* 36 (1968): 142-151.

    Argues that there is "a real and fundamental ambiguity,
    contradiction, or obscurity" in each of Maxwell's major
    discussions of electrostatics, but a different one in
    each. Finally, in the *Treatise*, Maxwell's equations
    mask his physical ideas. Displacement was initially
    introduced as a synonym for dielectric displacement.
    In later versions of the theory, however, Maxwell altered
    the signs appearing in his equations in such a way as
    to render those governing displacement incompatible
    with his earlier conception. As a result, displacement
    came to be seen by his successors as a conception
    original with him, and his intended meaning became lost.

1124. Buchwald, Jed Z., "William Thomson and the Mathematiza-
    tion of Faraday's Electrostatics." *Hist. Studs. Phys.
    Sci.* 8 (1977): 101-136.

    Argues that Thomson's application of Fourier's heat
    diffusion equation to electrostatics did not arise
    from an attempt to express Faraday's ideas about elec-
    tric force in mathematical terms. Rather, his primary
    purpose initially was to use Laplace's theory of attraction

to solve a problem in heat theory. Only subsequently, when he became aware of the difficulties associated with Poisson's conception of a physical layer of electric fluid on the surface of a conductor, did he abandon the fluid theory of electricity and seek to reconcile Green's analysis based on the notion of potential with Faraday's doctrines.

1125.   Buchwald, Jed Z., "The Hall Effect and Maxwellian Electrodynamics in the 1880's: Part I, The Discovery of a New Electric Field." *Centaurus* 23 (1979-80): 51-99; "Part II, The Unification of Theory, 1881-1893." *Ibid.*, pp. 118-162.

A detailed account of Hall's discovery based on his MS notebooks, and setting it securely within a tradition of electromagnetic theorizing, grounded on Maxwell's *Treatise*, in which electrical conduction was regarded as an entirely secondary field phenomenon. Shows, too, how the Hall effect was absorbed into the Maxwellian theory yet at the same time highlighted difficulties in the treatment of conductivity within that theory.

1126.   Buchwald, Jed Z., "The Abandonment of Maxwellian Electrodynamics: Joseph Larmor's Theory of the Electron." *Arch. Int. Hist. Sci.* 31 (1981): 135-180, 373-438.

A detailed technical account of the difficulties within Maxwellian electrodynamics arising from the notion that continuum theory could serve as a sufficient basis for electromagnetism, and of late 19th-century British responses to these difficulties. Argues that, as a result of Larmor's work, during the period 1894-97 "the most basic principles of Maxwell's theory of electromagnetism were abandoned, and the entire subject was reconstructed on a new foundation—the electron." Emphasizes "how completely foreign the idea of a current as a flow of charged particles was in Maxwellian theory, and therefore how very difficult it was for Larmor, independently of Lorentz, to achieve a theory that resembles Lorentz's in many respects."

*   Cantor, G.N., and M.J.S. Hodge, eds., *Conceptions of Ether: Studies in the History of Ether Theories, 1740-1900.*

Cited herein as item 238.

1126a. Cazenobe, Jean, "Comment Hertz a-t-il eu l'idée des
   ondes hertziennes?" *Rev. d. synthèse* 101 (1980):
   345-382.

   Exposes the background to Hertz' discovery of radio
   waves, stressing the remarkable interplay of theoretical
   and experimental components in his work and arguing that
   though the objective of the research appeared clear
   enough in retrospect, it was by no means so clear before-
   hand.

1127. Chalmers, A.F., "The Electromagnetic Theory of James
   Clerk Maxwell and some Aspects of Its Subsequent
   Development." Ph.D. thesis, University of London,
   1971.

   Provides the basis for the two papers listed below
   (items 1128, 1129). Includes in addition a chapter on
   "The Subsequent Extension of Maxwell's Theory," in
   which Helmholtz' theory of a polarizable aether, seen
   as a compromise between the action at a distance and
   continuous field theories, is presented as the crucial
   link between the work of Maxwell and that of Hertz and
   Lorentz.

1128. Chalmers, A.F., "The Limitations of Maxwell's Electromagnetic Theory." *Isis* 64 (1973): 469-483.

   Identifies several different limitations of Maxwell's
   theory, including (i) Maxwell's conception that light
   was a mechanical state of the aether, arising from a
   mechanical interaction between the matter of the source
   and the surrounding aether and not from *electrical* dis-
   turbances, meant that he failed to recognize the possibility of electromagnetic radiation; (ii) his conception of charge was vague and unsatisfactory; yet (iii)
   despite its vagueness, it was precise enough to lead
   to some falsifiable (and ultimately falsified) conclusions. These limitations are attributed to the theory's
   being, ironically, too much a theory about mechanisms in
   the aether, as a result of which it lacked important
   elements provided by the rival action-at-a-distance
   theory.

1129. Chalmers, A.F., "Maxwell's Methodology and his Application of it to Electromagnetism." *Studs. Hist. Phil.
   Sci.* 4 (1973): 107-164.

   Argues that Maxwell's innovations in electromagnetism
   were achieved in spite of the methodology to which he

purportedly subscribed: in particular, *contra* Duhem, Maxwell was led to the concept of displacement and hence the idea that light was an electromagnetic phenomenon, not by his various attempts to reduce electricity and magnetism to the principles of mechanics, but by arguments arising within the science of electricity itself. Analyzes a number of persisting (and connected) difficulties in Maxwell's notions of displacement and charge, arguing that Maxwell was actually hindered from resolving these by his methodological views.

1130.   Cranefield, Paul F., "Clerk Maxwell's Corrections to the Page Proofs of 'A Dynamical Theory of the Electromagnetic Field.'" *Ann. Sci.* 10 (1954): 359-362.

Maxwell's corrected page proofs, now in the Johns Hopkins University Library, show minor changes in scientific content in discussions of (i) the divergence of the vector potential and (ii) the optical properties of gold leaf.

1131.   Cuvaj, Camillo, "Henri Poincaré's Mathematical Contributions to Relativity and the Poincaré Stresses." *Amer. J. Phys.* 36 (1968): 1102-1113.

Gives a summary account of some of the main achievements in Poincaré's major paper of 1906 on the theory of the electron. An addendum to this paper with corrections is in *Amer. J. Phys.* 38 (1970): 774-775.

1132.   Doran, B.G., "Origins and Consolidation of Field Theory in Nineteenth-Century Britain: From the Mechanical to the Electromagnetic View of Nature." *Hist. Studs. Phys. Sci.* 6 (1975): 132-260.

A general and non-mathematical account in which Larmor's work in the 1890s is portrayed as the culmination of a long tradition in British physics of non-mechanical theories of the aether. The argument rests, however, on a number of dubious re-interpretations of earlier workers.

1133.   Duhem, Pierre, *Les théories électriques de J. Clerk Maxwell: Etude historique et critique.* Paris: A. Hermann, 1902. 238 pp.

A famous polemical critique of Maxwell's ideas.

1134.   Goldberg, Stanley, "The Abraham Theory of the Electron: The Symbiosis of Experiment and Theory." *Arch. Hist. Exact Sci.* 7 (1970): 7-25.

Shows how Abraham (1875-1922) attempted to derive a
wholly electrodynamic basis for mechanics on the assump-
tion of a rigid electron, Maxwell's equations, and an
absolute frame of reference determined by the aether.
Contrasts Abraham's views with Lorentz' theory based
on a deformable electron, assesses the relationship
between his work and Kaufmann's experimental investigations, and discusses the limitations of his approach.

1135. Goldberg, Stanley, "The Lorentz Theory of Electrons
       and Einstein's Theory of Relativity." *Amer. J. Phys.*
       37 (1969): 982-994.

Reviews the development of Lorentz' theory of electrons in so far as it relates to the electrodynamics
of moving bodies. Argues that the principle of relativity did not play an important role in Lorentz'
theory, and that though Lorentz eventually realized the
distinctions between his own work and that of Einstein,
he was unwilling to embrace Einstein's formulation completely and thereby to reject the aether.

1136. Goldberg, Stanley, "In Defense of Ether: The British
       Response to Einstein's Special Theory of Relativity,
       1905-1911." *Hist. Studs. Phys. Sci.* 2 (1970): 89-125.

Shows that in Britain in these years Einstein's
theory was largely neglected, and the concept of the
aether generally maintained. Attributes British slow-
ness to come to terms with Einstein to the fact that
most British theoreticians were trained at Cambridge,
and that through the Tripos examinations their training
was directed chiefly at questions of aether mechanics.

1137. Gooding, David, "Conceptual and Experimental Bases of
       Faraday's Denial of Electrostatic Action at a Dis-
       tance." *Studs. Hist. Phil. Sci.* 9 (1978): 117-149.

Argues *contra* T.M. Levere (item 811), L.P. Williams
(item 105), J. Agassi (item 29) and others that Faraday's
rejection of action at a distance is not inconsistent
with his theory of induction, and is derived from his
fundamental "principle of induction" which evolved
from his experiments in the years 1835-1838.

*   Gooding, David, "Metaphysics versus Measurement: The
       Conversion and Conservation of Force in Faraday's
       Physics."

Cited herein as item 973.

1138. Gooding, David, "Faraday, Thomson and the Concept of the
       Magnetic Field." *Brit. J. Hist. Sci.* 13 (1980):
       91-120.

       An account of the interactions during the 1840s be-
       tween Faraday and W. Thomson over the transmission of
       electrical and magnetic influences. Includes a detailed
       rebuttal of Doran's view (item 1132) that Thomson in-
       vented the field concept and guided Faraday toward
       field theory, arguing instead that Faraday had always
       preferred the field-theoretical approach, and that
       Thomson's role was to help Faraday draw the idea of a
       field out of his own representations of the phenomena.

1139. Gooding, David, "Final Steps to the Field Theory:
       Faraday's Study of Magnetic Phenomena, 1845-1850."
       Hist. Studs. Phys. Sci. 11 (1981): 231-275.

       Criticizes the supposition that Faraday had already
       by 1832 acquired the theory of magnetic lines of force,
       "the extension and confirmation of which was to be his
       life's work." Argues that, on the contrary, Faraday
       did not have a fully articulated theory of magnetism
       until the late 1840s, and describes the stages through
       which his thinking evolved at that period. Faraday's
       ideas eventually stabilized, it is concluded, "because
       he had at last forged a theoretical link between his
       deep assumptions and a wide range of natural phenomena."

1140. Gooding, David, "Empiricism in Practice: Teleology,
       Economy and Observation in Faraday's Physics." *Isis*
       73 (1982): 46-67.

       Explores in detail Faraday's work on magnetism, in
       an attempt to clarify the respective roles of normative
       and empirical factors in the construction of a scien-
       tific theory. Concludes that "Faraday's contribution
       to field physics was not the discovery of bald, novel
       'facts' described in a neutral observation language.
       Instead, his real contribution was the language of ob-
       servables based on axes of alignment and lines of force.
       This language suggested a new approach to the explana-
       tion of electric and magnetic phenomena...."

1141. Heimann, P.M., "Maxwell and the Modes of Consistent
       Representation." *Arch. Hist. Exact Sci.* 6 (1969-70):
       171-213.

       Argues that there was a "fundamental dichotomy" in
       Maxwell's thinking on electromagnetism, whereby on some

occasions he took lines of force to be the basic entities of the theory and on others he sought to reduce
these to states of polarization of particles of matter
and aether.

1142. Heimann, P.M., "Faraday's Theories of Matter and Elec-
          tricity." *Brit. J. Hist. Sci.* 5 (1970-71): 235-257.

Argues *contra* L.P. Williams (item 105), that prior to
Faraday's "Speculation touching Electric Conduction and
the Nature of Matter" of 1844 his thoughts do not show
any particular correspondence to Boscovich's atomic
theories, and that the ideas expressed in the "Speculation" are more characteristic of a British tradition
exemplified by Priestley's writings, and differ notably
from those of Boscovich.

1143. Heimann, P.M., "Maxwell, Hertz, and the Nature of Elec-
          tricity." *Isis* 62 (1971): 149-157.

Emphasizes the contradictions in Maxwell's discussions
of the nature of electricity in his *Treatise* and elsewhere, and argues that Hertz' desire to eliminate these
was fundamental to his reformulation of Maxwell's theory.

1144. Hermann, Armin, "Der Kraftbegriff bei Michael Faraday
          und seine historische Wurzel." *Wissenschaft, Wirt-
          schaft und Technik: Studien zur Geschichte*, ed. Karl-
          Heinz Manegold. Munich: Bruckmann, 1969, pp. 469-476.

Discusses the extent to which Faraday might have
been indebted for his conception of "force" to the
Romantics and to *Naturphilosophie*.

*   Hesse, Mary B., *Forces and Fields: The Concept of Action
          at a Distance in the History of Physics*.

Cited herein as item 248.

1145. Hesse, Mary B., "Logic of Discovery in Maxwell's Elec-
          tromagnetic Theory." *Foundations of Scientific
          Method: The Nineteenth Century*, ed. Ronald N. Giere
          and R.S. Westfall. Bloomington/London: Indiana Uni-
          versity Press, 1973, pp. 86-114.

Investigates Maxwell's explicit discussion of physical
methods and their application in his electromagnetic
theory. Argues that, in his mature theory, Maxwell
attempted to justify his introduction of the displacement current by a generalized method of induction and

analogy, and by no means regarded the idea as a hypothetical concept.

1146.   Hirosige, Tetu, "Electrodynamics before the Theory of
        Relativity, 1890-1905." *Japanese Studs. Hist. Sci.*
        5 (1966): 1-49.

        Presents three alternative streams of thought guiding
        electrodynamics during this period, namely: (i) Hertz'
        axiomatic approach, (ii) Larmor's aetherial dynamics,
        and (iii) Lorentz' and Wiechert's theory of electrons.
        Describes how Lorentz' theory came to be widely (but by
        no means universally) accepted shortly after 1900, and
        shows how this carried with it the view that the aether
        was merely the seat of the electromagnetic field, and
        not, after all, a mechanical substance.

1147.   Hirosige, Tetu, "Origins of Lorentz' Theory of Electrons
        and the Concept of the Electromagnetic Field." *Hist.*
        Studs. Phys. Sci. 1 (1969): 151-209.

        An excellent analysis of the evolution of Lorentz'
        ideas on electromagnetism, from his early work on an
        electromagnetic theory of optics based on an action-
        at-a-distance conception, up to his major presentation
        of the electron theory in 1895. Concentrates particularly
        on Lorentz' gradual conceptualization of the electro-
        magnetic field as a dynamical state of a stationary
        aether devoid of all mechanical qualities, rather than
        as a mechanical system in the aether as it was for
        Maxwell.

1148.   Hirosige, Tetu, "The Ether Problem, the Mechanistic
        World View, and the Origins of the Theory of Rela-
        tivity." *Hist. Studs. Phys. Sci.* 7 (1976): 3-82.

        A comprehensive survey of the 19th-century background
        to the work of Lorentz, Poincaré and Einstein on the
        "ether problem" and relativity. Argues that while both
        Lorentz and Poincaré were working within a traditional
        problem-situation which tried to reduce electromagnetism
        to mechanics (or vice-versa), Einstein was not. On the
        contrary, under the influence of Mach's critique of the
        mechanistic world-view--seen here as much more important
        for Einstein than his criticisms of absolute space and
        time--Einstein was seeking a unification of electromag-
        netism and mechanics "at a higher level" where the two
        theories were considered to be of equal standing.

*   Jolly, W.P., *Sir Oliver Lodge*.

        Cited herein as item 66.

*   Kargon, Robert H., "Model and Analogy in Victorian
        Science: Maxwell's Critique of the French Physicists."

        Cited herein as item 806.

1149.   Knudsen, Ole, "From Lord Kelvin's Notebook: Ether Specu-
        lations." *Centaurus* 16 (1971): 41-53.

        Publishes a brief extract from one of Kelvin's note-
        books, together with a brief commentary and explanatory
        notes. The passage in question is dated Jan. 6, 1859,
        and in it, Kelvin sets out his notion of the aether
        as an ideal elastic substance which he wishes to sub-
        stitute for the traditional conception of the aether
        as made up of discrete particles exerting forces on
        each other at a distance. The Faraday Effect figures
        prominently in the discussion.

1150.   Knudsen, Ole, "The Faraday Effect and Physical Theory."
        Arch. Hist. Exact Sci. 15 (1976): 235-281.

        Emphasizes the role played by the Faraday Effect in
        the development of Maxwell's electromagnetic theory,
        namely that it convinced him that the magnetic field
        was constituted of aetherial vortices. Argues that it
        was this conviction which lay behind Maxwell's well-
        known difficulties concerning the nature of electricity
        which, it is shown, vitiated even his detailed analysis
        of the Faraday Effect itself. Contrasts Maxwell's
        problems here with the power of the Continental action-
        at-a-distance approach as displayed in Carl Neumann's
        analysis of the same effect.

1151.   Knudsen, Ole, "Electric Displacement and the Development
        of Optics after Maxwell." *Centaurus* 22 (1978): 53-60.

        Argues that Maxwell's concept of displacement was of
        a real physical quantity representing the polarization
        of the aether and with no specific connection with
        matter. Gibbs in his papers on electromagnetic optics
        followed this, but Lorentz, following Helmholtz, regarded
        displacement as a composite with distinct components
        of polarization in both matter and the aether. The
        separation of the two helped pave the way for the elec-
        tron theory.

1152. Knudsen, Ole, "19th Century Views on Induction in Moving
       Conductors." *Centaurus* 24 (1980): 346-360.

       The author takes as his starting point the opening
       paragraph of Einstein's 1905 relativity paper containing
       his famous comments about an asymmetry in the classical
       treatment of electromagnetic induction. He shows that
       questions of symmetry and invariance in connection with
       electromagnetic induction had been discussed by a number
       of 19th-century authors, and points to a general feature
       of these discussions, namely their linking of the problem
       of induction with the question of the relationship be-
       tween the motions of matter and the aether.

1153. Kristenova, Dagmar, and Irena Seidlerova, "The Beginnings
       of Magnetooptics." *Acta historiae rerum naturalium
       necnon technicarum.* Czechoslovak Studies in the His-
       tory of Science. Special Issue 2, 2 (1966): 3-41.

       A valuable survey of the field. Covers the period
       from the discovery of the Faraday effect (1845) to the
       Zeeman effect (1895), including discussions of the con-
       tributions of Maxwell and Kerr.

1154. La Forgia, Mauro, "Michael Faraday ed i suoi biografi."
       Physis 20 (1978): 123-146.

       Examines various biographies of Faraday, suggesting
       problems involved in attempting to "adapt Faraday's
       case to a preconceived pattern."

1154a. La Forgia, Mauro, and Carlo Tarsitani, "Analogia, modelli
       e teoria dinamica nel contributo di Maxwell all'elet-
       tromagnetismo." *Physis* 23 (1981): 525-554.

       From a study of Maxwell's construction of electro-
       magnetic field theory, the authors attempt to identify
       the relation between his epistemological views, the
       method he used in his research, and the particular
       theory he developed. They stress the importance in
       his work of the notion of a "dynamical explanation"
       based on the priority of the energy concept.

1155. Levere, T.H., "Faraday, Matter and Natural Theology:
       Reflections on an Unpublished Mansucript." *Brit. J.
       Hist. Sci.* 4 (1968-69): 96-107.

       Publishes a MS of a "personal memoir" of 1844 wherein
       Faraday, in clarifying a concept of force-atoms, appeals
       to the idea of God having created a lawlike world.

1156. Lützen, J., "Heaviside's Operational Calculus and the Attempts to Rigorise It." *Arch. Hist. Exact Sci.* 21 (1979-80): 161-200.

Examines several examples of Heaviside's non-rigorous use of the basics of operational calculus in electrical calculations, showing his extensive reliance on physical intuition. Traces the efforts of later mathematicians to render these techniques more rigorous.

1157. McCormmach, Russell, "J.J. Thomson and the Structure of Light." *Brit. J. Hist. Sci.* 3 (1967): 362-387.

Describes Thomson's speculations concerning a discontinuous structure in the electromagnetic field, based upon the presumed discreteness of Faraday-style tubes of force and leading to the view that light was granular in character. Discusses the relationship between Thomson's ideas and Einstein's notion of light quanta, suggesting that even though Thomson resolutely opposed the quantum theory, the familiarity of British physicists with his own ideas about light helped reconcile them to it.

1158. McCormmach, Russell, "H.A. Lorentz and the Electromagnetic View of Nature." *Isis* 61 (1970): 458-497.

An authoritative study of Lorentz' development of the electron theory as the foundation of a universal purely electromagnetic physics. The inherently non-mechanical character of the theory is emphasized, as is the authoritative position it came to occupy in physics, especially in Germany, around 1900.

1159. McCormmach, Russell, "Einstein, Lorentz, and the Electron Theory." *Hist. Studs. Phys. Sci.* 2 (1970): 41-87.

An exceptionally clear account of the evolution of Einstein's thought during the first ten years of the 20th century. Emphasizes the central place of electrodynamics in the problem situations with which Einstein was concerned, and the importance of Lorentz' work to him even in those situations where he deviated most sharply from Lorentz' conceptions.

1160. McGuire, J.E., "Forces, Powers, Aethers, and Fields." *Methodological and Historical Essays in the Natural and Social Sciences*, ed. Robert S. Cohen and Marx W.

Wartofsky. (Boston Studies in the Philosophy of
Science, 14). Dordrecht/Boston: D. Reidel, 1974,
pp. 119-159.

An attempt to isolate "some of the turning points in
the history of the emergence of field concepts as a
prolegomenon to understanding the dynamics of conceptual
change involved." Includes a discussion (which draws
heavily on Heimann's work (item 1141) of the ideas of
Maxwell, and, much more briefly, an account of the ideas
of Poynting, J.J. Thomson, Larmor and Lorentz.

1161. Miller, Arthur I., "A Study of Henri Poincaré's 'Sur la
Dynamique de l'Electron.'" *Arch. Hist. Exact Sci.*
10 (1973): 207-328.

A detailed discussion, in its historical context, of
Poincaré's notable attempt to formulate a purely electromagnetic theory of a deformable electron. The paper
includes extended analyses of earlier theories of the
electron due to Abraham and Lorentz, and argues for the
influential nature of Poincaré's work despite its being
soon overtaken by Einstein's. Emphasizes Poincaré's
adherence to an electromagnetic world-picture in which
the principle of relativity was a law open to experimental verification.

1162. Miller, Arthur I., "On Lorentz's Methodology." *Brit.
J. Phil. Sci.* 25 (1974): 29-45.

Argues, in refutation of a theory of E. Zahar, that
the Lorentz contraction was specifically posited to
account for the Michelson-Morley experiment, that the
contraction was derived from Newton's addition law
of velocities, and that it was regarded as *ad hoc* until
it was shown to have wider powers in 1904, relying until
then entirely on a plausibility argument that ether-
matter interactions would cause a molecular contraction.

1163. Miller, Arthur I., *Albert Einstein's Special Theory of
Relativity: Emergence (1905) and Early Interpretation
(1905-1911).* Reading, Mass.: Addison-Wesley, 1981.
xxviii + 466 pp.

A "biography" of Einstein's 1905 relativity paper
which includes a long introductory chapter (pp. 11-
121), with considerable mathematical detail, on "Electrodynamics: 1890-1905." This surveys the theoretical
contributions of Lorentz, Poincaré and Abraham and
their interaction with the experimental work of Kaufmann.

1163a. Miller, Arthur I., "Unipolar Induction: A Case Study of
          the Interaction between Science and Technology." *Ann.
          Sci.* 38 (1981): 155-189.

    Discusses attempts during the 19th and 20th centuries
    to explain unipolar induction following Faraday's
    discovery of the phenomenon in 1832. Argues that these
    display striking national differences that influenced
    where the first large-scale unipolar dynamo was built.
    Also discusses the effect of unipolar induction on
    Einstein's thinking in relation to the special theory of
    relativity.

1164. Miller, John David, "Henry Augustus Rowland and his
          Electromagnetic Researches." Ph.D. dissertation,
          Oregon State University, 1970.

    Describes the work of Rowland (1848-1901) on magnetic
    distribution, charge convection, transverse electric
    current measurement, the ratio of units, the Ohm, and
    the production of electricity from the aether. Dis-
    cusses Rowland's relations with other scientists such
    as Maxwell, Helmholtz, and Hall, and his final un-
    completed work on the properties of the aether.

1165. Miller, John David, "Rowland and the Nature of Electric
          Currents." *Isis* 63 (1972): 5-27.

    An account, based on then recently-discovered manu-
    script sources, of Rowland's life-long concern with
    experimental investigations of the nature of electric
    currents, especially his efforts to demonstrate magnetic
    effects due to moving electric charges.

1166. Miller, John David, "Rowland's Magnetic Analogy to
          Ohm's Law." *Isis* 66 (1975): 230-241.

    Describes how Rowland attempted to translate into
    mathematical form Faraday's analogy between a magnet
    and its surrounding force field, and a voltaic battery
    immersed in water, arriving ultimately at the conclu-
    sion that magnetic induction is related to the magnetic
    potential between two points by a law exactly analogous
    to Ohm's law for electrical currents.

1167. Molella, Arthur Philip, "Philosophy and Nineteenth-
          Century German Electrodynamics: The Problem of Atomic
          Action at a Distance." Ph.D. thesis, Cornell Univer-
          sity, 1972. 263 pp.

Focuses on the interaction between Weber's theory of electrodynamic action at a distance between "atoms" of electricity and contemporary German philosophical atomism, especially as expounded by G.T. Fechner and J.F.C. Zöllner.

1168. Moyer, Donald Franklin, "Energy, Dynamics, Hidden Machinery: Rankine, Thomson and Tait, Maxwell." *Studs. Hist. Phil. Sci.* 8 (1977): 251-268.

Discusses the role of generalized equations of motion in Maxwell's electrodynamics (following earlier suggestions by Rankine and Thomson and Tait) as a method of generating mechanical explanations even though the underlying machinery remains hidden.

1169. Moyer, Donald Franklin, "Continuum Mechanics and Field Theory: Thomson and Maxwell." *Studs. Hist. Phil. Sci.* 9 (1978): 35-50.

Briefly discusses some 19th-century developments in continuum mechanics, especially those due to William Thomson, and shows how Maxwell used a generalized form of Thomson's line of reasoning to construct the electromagnetic theory set out in his *Treatise*.

1170. Pyenson, Lewis, "Physics in the Shadow of Mathematics: The Göttingen Electron-Theory Seminar of 1905." *Arch. Hist. Exact Sci.* 21 (1980): 55-89.

Describes the material studied at the seminar as a summary of immediately pre-Einsteinian electron theory--mainly the work of Hertz, Abraham, Schwarzschild, Sommerfeld and, above all, Lorentz--and argues that the failure of the group to resolve in a satisfactory manner the outstanding problems of electromagnetic theory was due to their overemphasizing the purely mathematical techniques involved, at the expense of physical theory.

1171. Rosenfeld, L., "The Velocity of Light and the Evolution of Electrodynamics." *Nuovo Cimento, Supplements, Series 10* 4 (1956): 1630-1669.

Discusses the developments leading up to Maxwell's identification of light as an electromagnetic phenomenon, and also the less familiar story of Lorenz' independently arriving by a different route at the same conclusion. Stresses the importance for both men of a dynamical conception of nature which within a few years was waning in significance.

1172.   Schaffner, Kenneth F., "The Lorentz Electron Theory
        and Relativity." *Amer. J. Phys.* 37 (1969): 498-513.

        Traces Lorentz' work on the electrodynamics of
        moving bodies from 1887 to 1909. Discusses the evolving
        role played in Lorentz' theory by the contraction hypo-
        thesis, and the type of support this enjoyed within the
        theory. Emphasizes the non-reciprocal character of the
        transformation equations in Lorentz' theory, in which
        the aether continues to provide a privileged reference
        frame.

1173.   Schaffner, Kenneth F., "Outline of a Logic of Compara-
        tive Theory Evaluation with Special Attention to
        Pre- and Post-Relativistic Electrodynamics." *His-
        torical and Philosophical Perspectives of Science*,
        ed. R.H. Stuewer. (Minnesota Studies in the Philo-
        sophy of Science, vol. 5.) Minneapolis: University
        of Minnesota Press, 1970, pp. 311-354.

        Uses the logic to assess the relative standing in
        1905 of Lorentz' and Einstein's electrodynamic theories,
        concluding that the former ranked higher in "theoretical
        context sufficiency" and the latter in simplicity con-
        siderations.

*   Schaffner, Kenneth F., ed., *Nineteenth-Century Aether
        Theories*.

        Cited herein as item 833.

1174.   Siegel, Daniel M., "Completeness as a Goal in Maxwell's
        Electromagnetic Theory." *Isis* 66 (1975): 361-368.

        Suggests that Maxwell's electromagnetic theory, and
        in particular his modification of Ampère's law by the
        inclusion of the displacement current, should be seen
        as "a shining example of systematic and goal-oriented
        theoretical endeavor rewarded," where the goal was
        theoretical completeness, that is, a theory which en-
        abled one to calculate effects "in the limiting cases
        where the known formulae are inapplicable."

1175.   Siegel, Daniel M., "Classical-Electromagnetic and
        Relativistic Approaches to the Problem of Nonintegral
        Atomic Masses." *Hist. Studs. Phys. Sci.* 9 (1978):
        323-360.

        Discusses the competing answers provided by classical
        electromagnetic theory (in which, following Lorentz,

all mass is regarded as electromagnetic) and relativity
theory, to the question of why atomic masses are not
integral multiples of the mass of the hydrogen atom.
These answers at first developed independently, but
from 1916 they interacted. Not until the 1930s did
the relativistic approach prevail. It is emphasized,
following McCormmach, that in the early days of relativity theory, Lorentz' theory "was not some hoary
predecessor, but rather a near contemporary, just as
new and full of promise."

1176. Simpson, Thomas K., "Maxwell and the Direct Experimental
       Test of His Electromagnetic Theory." *Isis* 57 (1966):
       411-432.

       Reports the results of an unsuccessful effort to
       trace any speculations of Maxwell's concerning the
       possibility of a direct experimental verification of
       his theory of electromagnetic propagation through a
       medium. Shows that many of the materials required for
       such direct experimentation were available in Maxwell's
       day, and takes his silence concerning them as evidence
       that his preoccupations were with different questions,
       and in particular with the nature of light rather than
       with electromagnetic phenomena for their own sake.

1177. Simpson, Thomas K., "A Critical Study of Maxwell's
       Dynamical Theory of the Electromagnetic Field in the
       Treatise on Electricity and Magnetism." Ph.D. thesis,
       Johns Hopkins University, 1968. 604 pp.

       Interprets Maxwell's *Treatise* as a systematic attempt
       to articulate Faraday's insights in the language of
       formal mathematical physics. Delineates two different
       phases in Maxwell's presentation of the dynamical
       theory in the *Treatise*, one inductive, the second
       deductive, with the dynamical theory proper appearing
       in the latter. This two-part form is seen to correspond
       to Maxwell's wider philosophy of science; the first
       phase is intended to uncover *a priori* intuitions of
       fundamental ideas from which the subsequent deductive
       phase can proceed.

1178. Simpson, Thomas K., "Some Observations on Maxwell's
       Treatise on Electricity and Magnetism." *Studs. Hist.
       Phil. Sci.* 1 (1970): 249-263.

       An article abstracted from the author's doctoral
       thesis (item 1177). Discusses the relationship between

Maxwell's metaphysics and his use of Lagrangian methods in developing his dynamical theory of the electro-magnetic field. Argues that, for Maxwell, "the Lagrangian mode of the dynamical theory is not ... simply a con-venience for an imperfect stage of a science; it is the appropriate mode for human knowledge of nature, which is essentially relative."

1178a. Sopka, Katherine R., "The Discovery of the Hall Effect: Edwin Hall's Hitherto Unpublished Account." *The Hall Effect and Its Applications*, ed. C.L. Chien and C.R. Westgate. New York/London: Plenum Press, 1980. pp. 523-545.

A transcript, with descriptive introduction, of Hall's notebook account of the 1879 investigation that led to the discovery of the transverse effect of a magnetic field on an electric current in a conductor.

1179. Spencer, J. Brookes, "On the Varieties of Nineteenth-Century Magneto-Optical Discovery." *Isis* 61 (1970): 34-51.

A brief account of magneto-optical work from Faraday to Zeeman, showing the interlinkage of theory and ex-periment and, more indirectly, the influence of Faraday in this field.

* Stein, Howard, "On the Notion of Field in Newton, Max-well and Beyond."

Cited herein as item 262.

1180. Susskind, Charles, "Observations of Electromagnetic-Wave Radiation before Hertz." *Isis* 55 (1964): 2-42.

Reviews various observations of electromagnetic-wave propagation between Maxwell and Hertz. These were almost all made by observers with a very limited back-ground in theoretical physics, and their significance was not recognized at the time.

1181. Topper, David Roy, "J.J. Thomson and Maxwell's Electro-magnetic Theory." Ph.D. thesis, Case Western Reserve University, 1970. 184 pp.

Describes Maxwell's work in electromagnetism, em-phasizing his use in his later papers of the Lagrangian formulation in order to ground the theory satisfactorily on dynamical principles while yet remaining ignorant of the actual mechanical systems presumed to be involved. Shows how Thomson extended this approach in order to show, through the use of cyclic co-ordinates, that

potential energy is formally equivalent to the kinetic
energy of hidden motions; and how on this basis he
banished forces from his dynamical theory in favour of
aetherial vortex motions and the like. Argues that
Thomson, far from upholding an electromagnetic theory
of matter, remained unfailingly committed to the
classical program of reducing electromagnetism to
dynamical principles.

1182. Topper, David Roy, "Commitment to Mechanism: J.J. Thom-
        son, the Early Years." *Arch. Hist. Exact Sci.* 7
        (1971): 393-410.

An article abstracted from the author's doctoral
thesis (item 1181). Argues that Thomson was committed
from the outset of his career to giving a complete
mechanical explanation of electromagnetic phenomena,
and shows how he accomplished this in a mathematical
proof involving the use of cyclic co-ordinates in
Lagrange's equations in order to reduce the potential
energy term (and hence the concept of force) to kinetic
energy, interpreted as the kinetic energy of hidden
motions.

1183. Tricker, R.A.R., *The Contributions of Faraday and Max-
        well to Electrical Science*. Oxford: Pergamon, 1966.
        xx + 289 pp.

Includes biographical sketches of both Faraday and
Maxwell, straightforward accounts of their work on
electromagnetism, and extracts from their chief writings
on the subject. Also has an interesting chapter on
"The Logical Status of the Law of Electromagnetic Induction."

1184. Turner, Joseph, "A Note on Maxwell's Interpretations
        of Some Attempts at Dynamical Explanation." *Ann. Sci.*
        11 (1955): 238-245.

Discusses the constraints that Maxwell imposed on the
development of dynamical explanations in physics, especially his requirements (i) that the mechanism
proposed be a "consistent representation" in the sense
that it is consistent with the fundamental principles
of dynamics, and (ii) that there be some independent
evidence for it.

1185. Turner, Joseph, "Maxwell on the Logic of Dynamical
        Explanation." *Phil. Sci.* 23 (1956): 36-47.

A brief but exceptionally clear exposition of Maxwell's attitude towards his theories, drawing for this
purpose the following useful distinctions based on

Maxwell's: (i) a *physical analogy* is a relation between a branch of one science and a branch of another, such that both branches possess the same mathematical form; (ii) a *dynamical analogy* is a physical analogy in which one of the branches of science involved is a branch of dynamics; (iii) a *dynamical explanation* is a dynamical analogy taken literally. In these terms, Maxwell's 1861 paper is seen as an attempt to provide a dynamical explanation of electromagnetism, while his 1864 paper has the more modest aim of providing a dynamical analogy.

1186. Turner, Joseph, "Maxwell on the Method of Physical Analogy." *Brit. J. Phil. Sci.* 6 (1956): 226-238.

An elaboration of Maxwell's views on the nature and usefulness of physical analogies.

1187. Wiederkehr, Karl Heinrich, "Wilhelm Webers Stellung in der Entwicklung der Elektrizitätslehre." Doctoral dissertation, Universität Hamburg, 1960. 254 pp.

A detailed account of Weber's career and scientific work. Emphasizes the importance for Weber of his early collaboration with Gauss.

*   Williams, L. Pearce, *Michael Faraday: A Biography.*

Cited herein as item 105.

1188. Williams, L. Pearce, *The Origins of Field Theory.* New York: Random House, 1966. xii + 148 pp.

Intended as a college text, this work draws heavily upon the same author's much larger intellectual biography of Faraday (item 105), but also devotes two chapters to setting the scientific background to Faraday's work and a concluding chapter to Maxwell's mathematization of Faraday's qualitatively expressed ideas. An excellent introduction to the subject.

1189. Wise, M. Norton, "The Flow Analogy to Electricity and Magnetism: Kelvin and Maxwell." Ph.D. thesis, Princeton University, 1977. 303 pp.

Focuses on the use by Kelvin and Maxwell (in his early work) of a flow analogy for electromagnetic forces in order to bring the mathematical techniques developed by Fourier to bear on Faraday's experimentally based conceptions. Presents on this basis an interpretation of the origins of electromagnetic field theory which emphasizes the role of mathematical techniques in producing conceptual change.

1190.   Wise, M. Norton, "The Mutual Embrace of Electricity and
        Magnetism." *Science* 203 (1979): 1310-1318.

        Explores the creative role of Maxwell's image of
        "mutually embracing curves" in the evolution of his
        electromagnetic theory, especially in the development
        of his ideas about the displacement current.

1191.   Wise, M. Norton, "The Flow Analogy to Electricity and
        Magnetism. Part I: William Thomson's Reformulation
        of Action at a Distance." *Arch. Hist. Exact Sci.*
        25 (1981-82): 19-70.

        Examines Thomson's work in terms of the developing
        character of British dynamical theories of mathematical
        physics, concentrating on its relationship to the work
        of Faraday and Maxwell.

1192.   Woodruff, A.E., "Action at a Distance in Nineteenth
        Century Electrodynamics." *Isis* 53 (1962): 439-459.

        Describes the developments leading up to Weber's
        formulation of his well-known expression for the force
        between two moving electric charges, and also the
        principal difficulties that others, especially Helm-
        holtz and Maxwell, found in Weber's conception.

1193.   Woodruff, A.E., "The Contributions of Hermann von Helm-
        holtz to Electrodynamics." *Isis* 59 (1968): 300-311.

        A good clear account which brings out the relationship
        of Helmholtz' work to that of both his predecessors
        (Neumann, Weber, Maxwell) and Hertz.

(iii) General

1194.   Bouthillon, Léon, "L'invention de la diode." *Rev.
        Hist. Sci.* 9 (1956): 354-356.

        Merely lists the major discoveries that mark the
        steps to the invention of the diode and names those
        associated with them from the 18th century to 1904.

1195.   [British Association for the Advancement of Science],
        *Reports of the Committee on Electrical Standards:
        A Record of the History of "Absolute Units" and of
        Lord Kelvin's Work in Connection with These*, ed.
        F.E. Smith. Cambridge: Cambridge University Press,
        1913. xxiv + 807 pp.

        A reprint of all the reports from 1862 to 1912 on
        the various attempts to determine absolute units of

resistance and current and the consideration of a variety of physical constants. Members of the Committee included Kelvin, Maxwell, Joule, Wheatstone and Fitz-Gerald.

* Cohen, I. Bernard, "Conservation and the Concept of Electric Charge: An Aspect of Philosophy in Relation to Physics in the Nineteenth Century."

Cited herein as item 791.

1196. Ekelöf, Stig, "The Genesis of the Wheatstone Bridge." *Technology and its Impact on Society: Tekniska Museet Symposium No. 1, 1977.* Stockholm: Tekniska Museet, 1979, pp. 81-93.

Considers the respective claims of Wheatstone and S.H. Christie to the invention of the "Wheatstone" Bridge.

1197. Finn, Bernard S., "Thermoelectricity." *Advances in Electronics and Electron Physics* 50 (1980): 176-240.

Traces the history of the investigation of thermo-electricity from the discovery of the Seebeck effect until the end of the 19th century. Includes discussion on solid state physics, low temperature physics, and the relationship between experiment and macroscopic (thermodynamical) and microscopic (electron) theories.

1198. Gorman, Mel, "Faraday on Lightning Rods." *Isis* 58 (1967): 96-98.

Describes three letters of Faraday's (1839-1845) in which he states his full support for the installation of lightning rods.

1199. Guralnick, Stanley M., "The Contents of Faraday's Electrochemical Laws." *Isis* 70 (1979): 59-75.

Describes the conceptual history of the laws, and how the meaning of the laws changed with developing concepts of electricity and matter--from Faraday's use of "force" as the fundamental notion, to the increasingly atomistic notions linked with other developing physical theories in the work of Joule, William Thomson, Maxwell, Helmholtz and others.

1200. Hopley, I.B., "Maxwell's Work on Electrical Resistance. I: The Determination of the Absolute Unit of Resistance." *Ann. Sci.* 13 (1957): 265-272.

Describes the procedures and apparatus used by Maxwell, Fleeming Jenkin and Balfour Stewart in 1863 to determine experimentally the absolute unit of resistance.

1201. Hopley, I.B., "Maxwell's Work on Electrical Resistance. II: Proposals for the Re-Determination of the B.A. Unit of 1863." *Ann. Sci.* 14 (1958): 197-210.

Describes the method and theory of three proposals by Maxwell to redetermine the unit of resistance. Two are in draft form and the "most ingenious" is reconstructed from two circuit diagrams. All involve measurement and calculation of the mutual inductance of two coils.

1202. Hopley, I.B., "Maxwell's Work on Electrical Resistance. III: Improvement on Mance's Method for the Measurement of Battery Resistance." *Ann. Sci.* 15 (1959): 51-55.

Prints an unpublished note of Maxwell's, with a short commentary on the efficacy of the method and a comparison of it with Lodge's later similar method.

1203. Hopley, I.B., "Maxwell's Determination of the Number of Electrostatic Units in One Electromagnetic Unit of Electricity." *Ann. Sci.* 15 (1959): 91-108.

Describes the method, the theory and the problems associated with Maxwell's determination of the ratio between absolute electrostatic and electromagnetic units, published in 1868. Also similarly describes two different later proposals for dealing with the same problem.

1203a. Jordan, D.W., "The Adoption of Self-Induction by Telephony, 1886-1889." *Ann. Sci.* 39 (1982): 433-461.

A careful analysis of the events surrounding the well-known controversy between Heaviside and W.H. Preece over the place of self-induction in telephone transmission and, ultimately, of mathematical (Maxwellian) theory in electrical engineering practice.

1203b. Jordan, D.W., "D.E. Hughes, Self-Induction and the Skin-Effect." *Centaurus* 26 (1982): 123-153.

Discusses the debate engendered by results on self-induction in conducting wires announced by Hughes in 1886 which challenged existing theoretical expectations

and led eventually to a general re-evaluation of the
status of mathematical theory in practical electrical
engineering.

1204. McLaughlin, P.J., *Nicholas Callan, Priest-Scientist*
(1799-1864). Dublin: Clonmore and Reynolds/London:
Burns and Oates, 1965. 128 pp.

Describes the life of the almost forgotten Irish
inventor of the induction coil, also renowned in his
time for his batteries and electromagnets. Subject
of an essay review by N.H. de V. Heathcote, *Ann. Sci.*
21 (1965): 145-167.

1205. Robotti, Nadia, "L'elettrone di Stoney." *Physis* 21
(1979): 103-143.

Describes Stoney's work from 1855-1910, concentrating
on his concept of unit electric charge--his electron--
and its relation to spectral theory, the interaction
of light and matter, and Faraday's laws.

1206. Rutenberg, D., "The Early History of the Potentiometer
System of Electrical Measurement." *Ann. Sci.* 4
(1939-40): 212-243.

Traces the development in theory and design of the
potentiometer, designed to measure electromotive force
unaffected by the usual polarizations occurring in
galvanic cells, from Poggendorff's first attempts
(1841) to the end of the 19th century.

1207. Stuewer, Roger H., "Non-Einsteinian Interpretations of
the Photoelectric Effect." *Historical and Philo-*
sophical Perspectives of Science, ed. R.H. Stuewer.
(Minnesota Studies in the Philosophy of Science,
vol. 5.) Minneapolis: University of Minnesota Press,
1970, pp. 246-263.

Looks at alternative, classical, explanations of the
photoelectric effect proposed by H.A. Lorentz, J.J.
Thomson, Arnold Sommerfeld and O.W. Richardson in the
period 1910-13 that attempted to question Einstein's
interpretation.

1208. Stuewer, Roger H., "Hertz's Discovery of the Photo-
electric Effect." *Proceedings, 13th International*
Congress for the History of Science, Moscow, 1971
(Moscow, 1974), vol. 6, pp. 35-43.

Describes Hertz's 1886-87 experiments.

1209. Tarsitani, Carlo, "La scopertà dell'effetto fotoelet-
       trico e il suo ruolo nello sviluppo della teoria
       quantistica: un caso storico di rapporto teoria-
       esperimento." *Physis* 20 (1978): 237-269.

       Includes a short section (pp. 237-244) on pre-electron
       investigations of the effect, 1887-99.

1210. Tunbridge, Paul A., "A Letter by William Thomson, F.R.S.,
       on the 'Thomson Effect.'" *Notes Rec. Roy. Soc. Lond.*
       26 (1971): 229-232.

       Publishes a letter of Thomson's to De la Rive prompted
       by a statement of his concerning the "electric convection
       of heat," establishing his own priority and also describing
       the means by which he made the discovery.

INDEX

Authors are indicated by an *.

Abbe, Ernst, 889, 890; *biog.* 31; *coll. works* 108
Abraham, Max, 1134, 1161, 1163, 1170
d'Abro, A., *232
Académie des Sciences de Paris (*see*: Paris, Académie des Sciences)
Academy of Sciences of the U.S.S.R. (*see*: St. Petersburg, Academy of Sciences)
Acloque, Paul, *911
Adams-Reilly, A., *86b
Adickes, Erich, *349b
Adie, A., 887
Aepinus, Franz Ulrich Theodosius, 346, 407, 594, *637, 637a, 676, 681
Agassi, Joseph, *28, *303, *351a, *780, *1111, 1137
d'Agostino, Salvo, *883a, *1112, *1113, *1114, *1115, *1116, *1117
Airy, G.B., 845, 927
Aiton, E.J., *532
d'Albe, E.E. Fournier, *29
d'Alembert, Jean Le Rond, 309, 323, 326, 351a, 394, 396, 543, 545, 548, 551, 552, 553, 553a, 556, 559, 561, 564, 568, 569, 570, 572
Algarotti, Francesco, 365, 415, 464
Amici, Giovan Battista, 888
Amontons, Guillaume, 625
Ampère, André-Marie, 234, 720, 741, 745, 746, 797, 817, 1081, 1082, 1083, 1085, 1086, 1087, 1088, 1090, 1094, 1096, 1099, 1104, 1105, 1106, 1107, 1108, 1110, 1174; *biog.* 72, 94; *coll. works* 109; *corresp.* 110
Andoyer, Henri, *710
Andrews, T., 1018
Ångström, A.J., 13
Apmann, Robert P., *912
Arago, Dominique François Jean, *1, *147, 714, 716, 718, 724, 727, 1099; *coll. works* 111
Archimedes, 586a
Arden, James, 381

293